A Random Tiling Model for
Two Dimensional Electrostatics

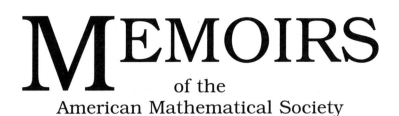

MEMOIRS
of the
American Mathematical Society

Number 839

A Random Tiling Model for Two Dimensional Electrostatics

Mihai Ciucu

November 2005 • Volume 178 • Number 839 (third of 5 numbers) • ISSN 0065-9266

American Mathematical Society
Providence, Rhode Island

2000 *Mathematics Subject Classification.*
Primary 82B23, 82D99 [Part A];05A15 [Part B];
Secondary 05A16, 41A63, 60F99 [Part A]; 82B23 [Part B].

Library of Congress Cataloging-in-Publication Data

Ciucu, Mihai, 1968–
    A random tiling model for two dimensional electrostatics / Mihai Ciucu.
        p. cm. — (Memoirs of the American Mathematical Society, ISSN 0065-9266 ; no. 839)
    "Volume 178, number 839 (third of 5 numbers)."
    Includes bibliographical references.
    ISBN 0-8218-3794-X (alk. paper)
    1. Tiling (Mathematics). 2. Electrostatics. 3. Statistical mechanics. I. Title. II. Series.

QA3.A57   no. 839
[QA166.8]
510 s—dc22
[537′.2]   2005050800

Memoirs of the American Mathematical Society

This journal is devoted entirely to research in pure and applied mathematics.

Subscription information. The 2005 subscription begins with volume 173 and consists of six mailings, each containing one or more numbers. Subscription prices for 2005 are $606 list, $485 institutional member. A late charge of 10% of the subscription price will be imposed on orders received from nonmembers after January 1 of the subscription year. Subscribers outside the United States and India must pay a postage surcharge of $31; subscribers in India must pay a postage surcharge of $43. Expedited delivery to destinations in North America $35; elsewhere $130. Each number may be ordered separately; *please specify number* when ordering an individual number. For prices and titles of recently released numbers, see the New Publications sections of the *Notices of the American Mathematical Society*.

Back number information. For back issues see the *AMS Catalog of Publications*.

Subscriptions and orders should be addressed to the American Mathematical Society, P. O. Box 845904, Boston, MA 02284-5904, USA. *All orders must be accompanied by payment.* Other correspondence should be addressed to 201 Charles Street, Providence, RI 02904-2294, USA.

Copying and reprinting. Individual readers of this publication, and nonprofit libraries acting for them, are permitted to make fair use of the material, such as to copy a chapter for use in teaching or research. Permission is granted to quote brief passages from this publication in reviews, provided the customary acknowledgment of the source is given.

Republication, systematic copying, or multiple reproduction of any material in this publication is permitted only under license from the American Mathematical Society. Requests for such permission should be addressed to the Acquisitions Department, American Mathematical Society, 201 Charles Street, Providence, Rhode Island 02904-2294, USA. Requests can also be made by e-mail to reprint-permission@ams.org.

Memoirs of the American Mathematical Society is published bimonthly (each volume consisting usually of more than one number) by the American Mathematical Society at 201 Charles Street, Providence, RI 02904-2294, USA. Periodicals postage paid at Providence, RI. Postmaster: Send address changes to Memoirs, American Mathematical Society, 201 Charles Street, Providence, RI 02904-2294, USA.

Dedicated to Professor Richard Stanley on the occasion of his sixtieth birthday

Contents

Abstract

The two parts of this Memoir contain two separate but closely related papers. In the paper in Part A we study the correlation of holes in random lozenge (i.e., unit rhombus) tilings of the triangular lattice. More precisely, we analyze the joint correlation of these triangular holes when their complement is tiled uniformly at random by lozenges. We determine the asymptotics of the joint correlation (for large separations between the holes) in the case when one of the holes has side 1, all remaining holes have side 2, and the holes are distributed symmetrically with respect to a symmetry axis. Our result has a striking physical interpretation. If we regard the holes as electrical charges, with charge equal to the difference between the number of down-pointing and up-pointing unit triangles in a hole, the logarithm of the joint correlation behaves exactly like the electrostatic potential energy of this two-dimensional electrostatic system: it is obtained by a Superposition Principle from the interaction of all pairs, and the pair interactions are according to Coulomb's law. The starting point of the proof is a pair of exact lozenge tiling enumeration results for certain regions on the triangular lattice, presented in the second paper.

The paper in Part B was originally motivated by the desire to find a multi-parameter deformation of MacMahon's simple product formula for the number of plane partitions contained in a given box. By a simple bijection, this formula also enumerates lozenge tilings of hexagons of side-lengths a, b, c, a, b, c (in cyclic order) and angles of 120 degrees. We present a generalization in the case $b = c$ by giving simple product formulas enumerating lozenge tilings of regions obtained from a hexagon of side-lengths $a, b+k, b, a+k, b, b+k$ (where k is an arbitrary non-negative integer) and angles of 120 degrees by removing certain triangular regions along its symmetry axis. The paper in Part A uses these formulas to deduce that in the scaling limit the correlation of the holes is governed by two dimensional electrostatics.

2000 Mathematics Subject Classification Numbers, paper in Part A: Primary 82B23, 82D99; Secondary 05A16, 41A63, 60F99.

Keywords: dimer model, random tilings, lozenge tilings, perfect matchings, exact enumeration, asymptotic enumeration, correlation, scaling limit, electrostatics.

2000 Mathematics Subject Classification Numbers, paper in Part B: Primary 05A15; Secondary 82B23.

Keywords: plane partitions, exact enumeration, simple product formulas, lozenge tilings, perfect matchings.

Received by the editor August 23, 2003; and in revised form on July 11, 2004 and October 11, 2004.

The research for the paper in Part A was supported in part by NSF grants DMS 9802390 and DMS 0100950. The research for the paper in Part B was supported by a Membership at the Institute for Advanced Study.

Part A

A Random Tiling Model for Two Dimensional Electrostatics

1

Introduction

Monomer-monomer and especially dimer-dimer correlations[1] on a plane bipartite lattice (especially the square and hexagonal lattice) have been studied quite extensively (see for instance [**12**], [**17**], [**20**] and [**21**]). Color the vertices of the lattice white and black so that each edge has one white and one black endpoint. From the point of view of this paper, there is a fundamental difference between studying dimer-dimer and monomer-monomer correlations: the former have the same number of vertices of each color, while the latter have an excess of either a white or a black vertex.

In this paper we consider correlations of triangular holes on the hexagonal lattice.

This type of hole has the convenient feature that the difference between the number of its white and black constituent vertices is equal to the length of its side. We will be lead by our results to interpret the white vertices as elementary negative charges ("electrons"), and the black vertices as elementary positive charges ("positrons"), so that the triangular plurimers become charges of magnitude given by their side-length, and sign given by their orientation (up-pointing or down-pointing). The main result of this paper, Theorem 2.1 (see also its much simpler restatement (2.6)), implies that in the fairly general situation in which it applies (namely, when the holes are symmetrically distributed about an axis, and all holes have side 2, except for one of side 1, on the symmetry axis) the logarithm of the joint correlation of triangular holes behaves exactly like the two-dimensional electrostatic potential energy of the corresponding system of charges: it is obtained by a Superposition Principle from the interaction of all pairs, and the pair interactions are according to Coulomb's law. (It is now clear why dimer-dimer and monomer-monomer correlations are fundamentally different: dimers are neutral!)

To present our results in the background of previous related results in the literature, we point out the following facts.

First, there seem to be very few rigorous results in the literature on the asymptotics of the correlation of holes that are not unions of edges. The only ones the author is aware of are [**12**], [**17**], and [**9**] (they all deal with instances of two monomers; the first and third paper present explicit conjectures, while the second proves a special case of a conjecture in the first). It was the work of Fisher and Stephenson [**12**] that provided our original motivation for studying joint correlations of holes.

Second, we mention that there is an alternative approach for expressing joint correlations of holes on the hexagonal (or square) lattice due to Kenyon ([**21**, Theorem 2.3]; see also [**20**]). When it applies, it provides an expression for the joint

---

[1]The monomers, respectively the dimers, are interacting via a sea of dimers that cover all lattice sites not occupied by them.

correlation as a $k \times k$ determinant (where $2k$ is the total number of vertices in the holes). However, while not requiring symmetry, the set-up of Kenyon's approach limits its applicability to the case when all holes have even side (thus not accommodating our hole of side 1 on the symmetry axis), and, more restrictively, to the case when the total "charge" of the holes is zero[2]. The main advantage of our approach is that it sets no restriction on the the total charge.

Furthermore, our result (14.9) gives the joint correlation of a general distribution of collinear *monomers* on the square lattice, a situation in which none of the holes satisfies the even-sidedness required by Kenyon's approach.

Third, there are other discrete models in the physics literature (see e.g. the survey [27] by Nienhuis) believed to behave like a Coulomb gas. However, there are several important differences between them and our model: (1) we do not require that the total charge is 0, a fact built into the definition of the Coulomb gas model in the survey by Nienhuis [(2.7), 27]; (2) by studying correlation of holes, our discrete model seems quite different from the others in the literature; (3) for those models surveyed in [27] for which the (believed) Coulomb behavior is only asymptotic (as it is the case for our model), the arguments for their equivalence with the Coulomb gas model are only heuristic, while our results are proved rigorously; (4) in our model all states have the same energy, so the emergence of the Coulomb interaction is entirely due to the number of different geometrical configurations (unit rhombus tilings) compatible with the holes—in the language of physicists, our model is *stabilized by entropy*; by contrast, in all models surveyed in [27] different configurations have different energies, specified by a Hamiltonian—they are *stabilized by energy*. Furthermore, in our model the electrical charge has a purely geometric origin: the charges are holes in the lattice, and their magnitude is the difference between the number of right-pointing and left-pointing unit triangles in the hole.

Fourth, in a recent paper [22] Krauth and Moessner study numerically (using Monte Carlo algorithms) a very special case of the problem considered in this paper, namely just the two-point correlations of monomers on the square lattice. Their data leads them to conjecture that the two-point correlations on monomers behave like a Coulomb potential (the case of monomers on different sublattices appeared already in [12]; the new part is the simulation data for monomers on the same sublattice). Krauth and Moessner also state that they could not find these correlations worked out in the literature. Since in the current paper we address the case of $(2m + 2n + 1)$-point correlations, showing that they satisfy the much more general Superposition Principle, their remark suggests that our results are also new.

And fifth, shortly after the current paper was posted on the preprint archive (web address arxiv.org/abs/math-ph/0303067, March 2003), physicists D. A. Huse, W. Krauth, R. Moessner and S. L. Sondhi posted an article (arxiv.org/abs/cond-mat/0305318, May 2003) presenting numerical simulations that suggest a positive

---

[2]Indeed, Kenyon [20] defines the correlation of holes as the limit of $M(\hat{H}_{m,n})/M(H_{m,n})$ when $m, n \to \infty$, where $H_{m,n}$ is a toroidal hexagonal grid graph, and $\hat{H}_{m,n}$ is its subgraph obtained by removing the holes ($M(G)$ denotes the number of perfect matchings of the graph G). In order for $M(H_{m,n})$ to be non-zero one must have the same number of vertices in the two bipartition classes of the vertices of $H_{m,n}$. Thus, since $M(\hat{H}_{m,n})$ also has to be non-zero in order for Kenyon's correlation to be non-zero, the number of vertices in the two bipartition classes that fall in the holes must also be the same; i.e., the total charge of the holes must be zero.

answer to Question 15.1 (which concerns a three dimensional analog of the model presented in this paper) for the case of two monomers.

Definitions, statement of results
and physical interpretation

We will find it more convenient to present our results on dimer coverings of portions of the hexagonal lattice in the equivalent, dual language of lozenge tilings of lattice regions on the triangular lattice. In this language a monomer is just a unit triangle of the lattice. A lattice triangle of side two will be called a *quadromer*. The results proved in this paper involve only monomers and quadromers. More generally, triangular plurimers are lattice triangles of arbitrary side. A dimer becomes a lozenge—a unit rhombus covering precisely two unit triangles. We will often refer to lozenges as dimers.

Let m and n, and N be nonnegative integers. Consider also the nonnegative integers R_i, v_i, $i = 1, \ldots, m$ and R'_i, v'_i, $i = 1, \ldots, n$. Define the region

$$(2.1) \qquad H_N \begin{pmatrix} R_1 & \cdots & R_m & R'_1 & \cdots & R'_n \\ v_1 & & v_m & v'_1 & & v'_n \end{pmatrix}$$

as follows.

Consider a lattice hexagon H whose sides alternate between $2N + 4n + 1$ and $2N + 4m$, starting with the base. Denote by ℓ the vertical symmetry axis of H. Let u be the up-pointing unit triangle on ℓ whose base is $(N + 2m)\sqrt{3}$ units above the base of H (that is, the base of u is along the horizontal diagonal of H).

Let $D(R, v)$ be the down-pointing quadromer (i.e., down-pointing lattice triangle of side 2) whose base is centered R units to the left of ℓ, and $(2v + 1)\sqrt{3}/2$ units below the base of u (two instances of this appear in Figure 2.1). Let $U(R, v)$ be the up-pointing quadromer whose base is centered R units to the left of ℓ and $(2v + 1)\sqrt{3}/2$ units *above* the base of u (see Figure 2.1 for three instances of this).

Finally, let $\bar{D}(R, v)$ and $\bar{U}(R, v)$ be the mirror images in ℓ of the above quadromers. We define $H_N \begin{pmatrix} R_1 & \cdots & R_m & R'_1 & \cdots & R'_n \\ v_1 & & v_m & v'_1 & & v'_n \end{pmatrix}$ to be the region obtained from the hexagon H by removing u (a monomer), $D(R_i, v_i)$, $\bar{D}(R_i, v_i)$, $i = 1, \ldots, m$ and $U(R'_j, v'_j)$, $\bar{U}(R'_j, v'_j)$, $j = 1, \ldots, n$ (a total of $2m + 2n + 1$ holes).

We assume throughout the paper that $R_i \geq 1$, $i = 1, \ldots, m$, and $R'_j \geq 1$, $j = 1, \ldots, n$, so that ℓ separates the original plurimers from their mirror images. Figure 2.1 shows the region $H_2 \begin{pmatrix} 5 & 2 & 4 & 2 & 3 \\ 0 & 1 & 1 & 2 & 4 \end{pmatrix}$.

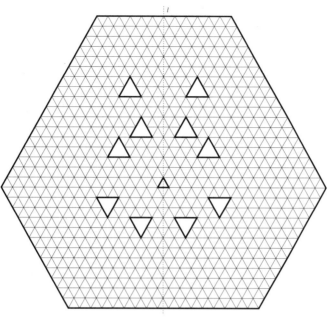

FIGURE 2.1. $H_2 \begin{pmatrix} 5 & 2 & 4 & 2 & 3 \\ 0 & 1 & 1 & 2 & 4 \end{pmatrix}$.

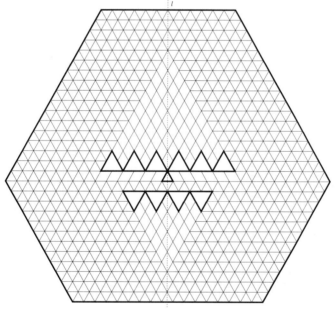

FIGURE 2.2. $H_2 \begin{pmatrix} 1 & 3 & 1 & 3 & 5 \\ 0 & 0 & 0 & 0 & 0 \end{pmatrix}$.

Define the *correlation at the center* (or simply correlation) of these $2m + 2n + 1$ plurimers by

(2.2)

$$\omega \begin{pmatrix} R_1 & \dots & R_m & R'_1 & \dots & R'_n \\ v_1 & & v_m & v'_1 & & v'_n \end{pmatrix} :=$$

$$\lim_{N \to \infty} \frac{\mathrm{M} \left(H_N \begin{pmatrix} R_1 & R_2 & \dots & R_m & R'_1 & R'_2 & \dots & R'_n \\ v_1 & v_2 & & v_m & v'_1 & v'_2 & & v'_n \end{pmatrix} \right)}{\mathrm{M} \left(H_N \begin{pmatrix} 1 & 3 & \dots & 2m-1 & 1 & 3 & \dots & 2n-1 \\ 0 & 0 & & 0 & 0 & 0 & & 0 \end{pmatrix} \right)},$$

where $\mathrm{M}(D)$ is the number of dimer coverings (i.e, lozenge tilings—tilings by unit rhombuses each covering precisely two unit triangles) of the lattice region D. The existence of the limit on the right hand side of (2.2) follows from Proposition 3.2.

In the above definition, the region at the denominator is obtained from the one in the numerator by packing the quadromers as tightly as possible around u—for the example of Figure 2.1 this is illustrated in Figure 2.2. As indicated by the latter figure, because of forced lozenges, the situation is equivalent to removing just two large plurimers from H, one down-pointing of side $2m$ and one up-pointing of side $2n + 1$.

Assume that the midpoints of the bases of the quadromers have coordinates

(2.3)

$$\begin{aligned} R_i &= A_i R \\ v_i &= q_i R_i + c_i = q_i A_i R + c_i \\ R'_j &= B_j R \\ v'_j &= q'_j R'_j + c'_j = q'_j B_j R + c'_j, \end{aligned}$$

where: $A_i > 0$, $q_i > 0$, $i = 1, \dots, m$ are fixed rational numbers chosen so that the pairs $(A_1, q_1), \dots, (A_m, q_m)$ are distinct; $B_j > 0$, $q'_j > 0$, $j = 1, \dots, n$ are fixed rational numbers so that the pairs $(B_1, q'_1), \dots, (B_n, q'_n)$ are distinct; $c_i \geq 0$, $i = 1, \dots, m$ and $c'_j \geq 0$, $j = 1, \dots, n$ are fixed integers; and R is an integer parameter.

The main result of this paper is the following.

THEOREM 2.1. *If the coordinates of the quadromers are as in (2.3) and $R \to \infty$, their joint correlation with the fixed monomer u is given asymptotically by*

(2.4)

$$\omega \begin{pmatrix} R_1 & \dots & R_m & R'_1 & \dots & R'_n \\ v_1 & & v_m & v'_1 & & v'_n \end{pmatrix} =$$

$$c_{2m,2n} \prod_{i=1}^{m} (2R_i)^2 \prod_{j=1}^{n} (2R'_j)^2 \frac{\prod_{j=1}^{n} (R'_j)^2 + 3(v'_j)^2}{\prod_{i=1}^{m} R_i^2 + 3v_i^2}$$

$$\times \prod_{1 \leq i < j \leq m} [(R_j - R_i)^2 + 3(v_j - v_i)^2]^2 \, [(R_j + R_i)^2 + 3(v_j - v_i)^2]^2$$

$$\times \frac{\prod_{1 \leq i < j \leq n} [(R'_j - R'_i)^2 + 3(v'_j - v'_i)^2]^2 \, [(R'_j + R'_i)^2 + 3(v'_j - v'_i)^2]^2}{\prod_{i=1}^{m} \prod_{j=1}^{n} [(R'_j - R_i)^2 + 3(v'_j + v_i)^2]^2 \, [(R'_j + R_i)^2 + 3(v'_j + v_i)^2]^2}$$

$$+ O(R^{4(m-n)^2 - 4m - 1}),$$

where

$$(2.5) \qquad c_{k,l} = \frac{2^{k+l}3^{k-(k-l)^2/2}}{\pi^{k+l}} \left(\prod_{j=0}^{k-1} \frac{(2)_j}{(1)_j(3/2)_j} \prod_{j=0}^{l-1} \frac{(j+2)_k}{(3/2)_j} \right)^2.$$

Remarkably, each factor in (2.4), except for the multiplicative constant $c_{2m,2n}$, is exactly equal to the Euclidean distance between the midpoints of the bases[3] of some pair of our $2m + 2n + 1$ plurimers (one monomer and $2m + 2n$ quadromers). Indeed:

- (i) $2R_i$ is the distance between $D(R_i, v_i)$ and $\bar{D}(R_i, v_i)$;
- (ii) $2R'_j$ is the distance between $U(R'_j, v'_j)$ and $\bar{U}(R'_j, v'_j)$;
- (iii) $[(R'_j)^2 + 3(v'_j)^2]^{1/2}$ is both the distance between $U(R'_j, v'_j)$ and monomer u, and the distance between $\bar{U}(R'_j, v'_j)$ and monomer u;
- (iv) $[R_i^2 + 3v_i^2]^{1/2}$ is both the distance between $D(R_i, v_i)$ and monomer u, and the distance between $\bar{D}(R_i, v_i)$ and monomer u;
- (v) $[(R_j - R_i)^2 + 3(v_j - v_i)^2]^{1/2}$ is both the distance between $D(R_i, v_i)$ and $D(R_j, v_j)$, and the distance between $\bar{D}(R_i, v_i)$ and $\bar{D}(R_j, v_j)$;
- (vi) $[(R_j + R_i)^2 + 3(v_j - v_i)^2]^{1/2}$ is both the distance between $D(R_i, v_i)$ and $\bar{D}(R_j, v_j)$, and the distance between $\bar{D}(R_i, v_i)$ and $D(R_j, v_j)$;
- (vii) $[(R'_j - R'_i)^2 + 3(v'_j - v'_i)^2]^{1/2}$ is both the distance between $U(R'_i, v'_i)$ and $U(R'_j, v'_j)$, and the distance between $\bar{U}(R'_i, v'_i)$ and $\bar{U}(R'_j, v'_j)$;
- $(viii)$ $[(R'_j + R'_i)^2 + 3(v'_j - v'_i)^2]^{1/2}$ is both the distance between $U(R'_i, v'_i)$ and $\bar{U}(R'_j, v'_j)$, and the distance between $\bar{U}(R'_i, v'_i)$ and $U(R'_j, v'_j)$;
- (ix) $[(R'_j - R_i)^2 + 3(v'_j + v_i)^2]^{1/2}$ is both the distance between $D(R_i, v_i)$ and $U(R'_j, v'_j)$, and the distance between $\bar{D}(R_i, v_i)$ and $\bar{U}(R'_j, v'_j)$;
- (x) $[(R'_j + R_i)^2 + 3(v'_j + v_i)^2]^{1/2}$ is both the distance between $D(R_i, v_i)$ and $\bar{U}(R'_j, v'_j)$, and the distance between $\bar{D}(R_i, v_i)$ and $U(R'_j, v'_j)$.

This remarkable phenomenon goes even further: if one defines the *charge* $\mathrm{ch}(Q)$ of a hole to be the number of its up-pointing unit lattice triangles minus the number of its down-pointing ones (this clearly gives 1 for the monomer u, and -2 and 2 for the quadromers of type D and U, respectively), the exponent with which each such factor occurs in (2.4) is precisely half the product of the charges of the corresponding two plurimers. Therefore, the statement of Theorem 2.1 can be rewritten as follows.

THEOREM 2.1 (EQUIVALENT RESTATEMENT). *Denote the* $(2m + 2n + 1)$ *triangular plurimers removed in region* (2.1) *by* $Q_1, \ldots, Q_{2m+2n+1}$, *and define the charge* $\mathrm{ch}(Q)$ *of* Q *to be the number of up-pointing unit lattice triangles of* Q *minus the number of its down-pointing unit triangles. Let* $\mathrm{d}(Q_i, Q_j)$ *denote the Euclidean distance between the midpoints of the bases of plurimers* Q_i *and* Q_j.

Then as the coordinates of the midpoints of the bases of the $2m+2n$ *quadromers approach infinity as specified by* (2.4), *the asymptotics of their joint correlation with*

---

[3]For the monomer u—involved in the factors of the fraction on the second line of (2.4)—instead of the midpoint of its base we need to consider the closest lattice point above it or below it, according as the factor in question is at the numerator or denominator, respectively.

the monomer u (which by definition is kept fixed at the origin) is given by

$$\omega(Q_1,\ldots,Q_{2m+2n+1}) = c_{2m,2n} \prod_{1\le i<j\le 2m+2n+1} \mathrm{d}(Q_i,Q_j)^{\mathrm{ch}(Q_i)\,\mathrm{ch}(Q_j)/2}$$

$$(2.6) \hspace{5cm} +O(R^{4(m-n)^2-4m-1}),$$

where the constant $c_{k,l}$ is given by (2.5).

Our result has the following striking physical interpretation.

Take the logarithm of both sides of (2.6), and assume $4(m-n)^2 - 4m \neq 0$. Since then the main term on the right hand side of (2.4) approaches either 0 or infinity as $R \to \infty$, the resulting contribution from the multiplicative constant is negligible. We obtain

$$(2.7) \quad \log\omega(Q_1,\ldots,Q_{2m+2n+1}) \sim \sum_{1\le i<j\le 2m+2n+1} \frac{\mathrm{ch}(Q_i)\,\mathrm{ch}(Q_j)}{2} \log \mathrm{d}(Q_i,Q_j),$$

where $\mathrm{d}(Q_i,Q_j)$ is the Euclidean distance between the plurimers Q_i and Q_j.

But the logarithm of the distance is, up to a negative multiplicative constant, just the potential in two dimensional electrostatics! So if we view the plurimers as *electrical charges* of (signed) magnitude given by the operator ch (in this case, just their side-length if they point upward, or minus their side-length if they point downward), the logarithm of the joint correlation (multiplied by a negative constant) acts precisely like the electrostatic potential energy of this two dimensional electrostatic system: it is obtained by a Superposition Principle from the interaction of all pairs, and the pair interactions are according to Coulomb's law.

Both for our proof of Theorem 2.1 and from the point of view of physical interpretation we find it useful to define another kind of plurimer correlation, in which the boundary of the defining regions makes its presence felt. To this end, define another family of regions as follows.

Cut the region $H_N \left(\begin{matrix} R_1 & \ldots & R_m & ; & R'_1 & & R'_n \\ v_1 & & v_m & ; & v'_1 & & v'_n \end{matrix} \right)$ along a zig-zag line that starts at the midpoint of its top side, follows ℓ as closely as possible on its right until it reaches the central monomer u, crosses ℓ along the western edge of u and then follows ℓ as closely as possible on its *left*, ending at the midpoint of the base (this cut is pictured in Figure 2.3). We define

$$W_N \left(\begin{matrix} R_1 & \ldots & R_m & ; & R'_1 & & R'_n \\ v_1 & & v_m & ; & v'_1 & & v'_n \end{matrix} \right)$$

to be the region obtained this way to the west of our cut, in which in addition the $N+2n$ lozenge positions above u and immediately to the left of the cut are distinguished and given weight $1/2$ (i.e., each lozenge tiling T of the regions W_N is weighted by $1/2^k$, where k is the number of distinguished lozenge positions occupied by a lozenge in T). An example is illustrated in Figure 2.3 (the distinguished lozenge positions are marked by shaded ellipses).

When $N \to \infty$ and the coordinates R_i, v_i, R'_i, v'_i of the quadromers are fixed, they maintain a fixed relative position with respect to the right boundary of the

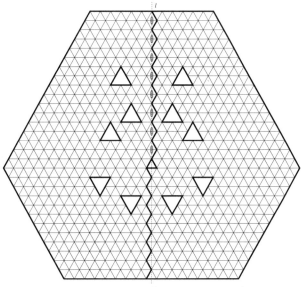

$$\text{FIGURE 2.3. } W_2 \begin{pmatrix} 5 & 2 & 4 & 2 & 3 \\ 0 & 1 & 1 & 2 & 4 \end{pmatrix}.$$

regions W_N. We define the *boundary-influenced correlation* ω_b by

$$\omega_b \begin{pmatrix} R_1 & \cdots & R_m & R'_1 & \cdots & R'_n \\ v_1 & \cdots & v_m & v'_1 & \cdots & v'_n \end{pmatrix} :=$$

(2.8)
$$\lim_{N \to \infty} \frac{\mathrm{M}\left(W_N \begin{pmatrix} R_1 & R_2 & \cdots & R_m & R'_1 & R'_2 & \cdots & R'_n \\ v_1 & v_2 & & v_m & v'_1 & v'_2 & & v'_n \end{pmatrix} \right)}{\mathrm{M}\left(W_N \begin{pmatrix} 1 & 3 & \cdots & 2m-1 & 1 & 3 & \cdots & 2n-1 \\ 0 & 0 & & 0 & 0 & 0 & & 0 \end{pmatrix} \right)}$$

(the fact that this limit exists follows by Lemma 5.1).

An important part of the proof of Theorem 2.1—and a result interesting on its own—is the determination of the asymptotics of ω_b.

THEOREM 2.2. *If the coordinates of the quadromers* $D(R_i, v_i)$, $i = 1, \ldots, m$ *and* $U(R_j, v_j)$, $j = 1, \ldots, n$ *are given by (2.3), the large R asymptotics of their boundary-influenced correlation is*

$$\omega_b \begin{pmatrix} R_1 & \cdots & R_m & R'_1 & \cdots & R'_n \\ v_1 & \cdots & v_m & v'_1 & \cdots & v'_n \end{pmatrix} =$$

$$\phi_{2m,2n} \prod_{i=1}^{m}(2R_i) \prod_{j=1}^{n}(2R'_j) \frac{\prod_{j=1}^{n}[(R'_j)^2 + 3(v'_j)^2]^{1/2}}{\prod_{i=1}^{m}[R_i^2 + 3v_i^2]^{1/2}}$$

$$\times \prod_{1 \le i < j \le m} [(R_j - R_i)^2 + 3(v_j - v_i)^2]\,[(R_j + R_i)^2 + 3(v_j - v_i)^2]$$

$$\times \frac{\prod_{1 \le i < j \le n}[(R'_j - R'_i)^2 + 3(v'_j - v'_i)^2]\,[(R'_j + R'_i)^2 + 3(v'_j - v'_i)^2]}{\prod_{i=1}^{m}\prod_{j=1}^{n}[(R'_j - R_i)^2 + 3(v'_j + v_i)^2]\,[(R'_j + R_i)^2 + 3(v'_j + v_i)^2]}$$

(2.9)
$$+ O(R^{2(m-n)^2 - 2m - 1}),$$

where

$$(2.10) \qquad \phi_{k,l} = \frac{2^k 3^{(k+l)/4 - (k-l)^2/4}}{\pi^{(k+l)/2}} \prod_{j=0}^{k-1} \frac{(2)_j}{(1)_j (3/2)_j} \prod_{j=0}^{l-1} \frac{(j+2)_k}{(3/2)_j}.$$

This result also has an interesting physical interpretation. Relabel the $m + n$ quadromers by $Q_1, \ldots, Q_{m+n}$. Assuming $2(m-n)^2 - 2m \neq 0$, one deduces from (2.9)—just as (2.7) was deduced from (2.4)—that

$$(2.11) \qquad \log \omega_b(Q_1, \ldots, Q_{m+n}) \sim \sum_{Q, Q' \in \mathcal{Q}, Q \neq Q'} \frac{\mathrm{ch}(Q) \, \mathrm{ch}(Q')}{4} \log \mathrm{d}(Q, Q'),$$

where ch denotes the charge, d the Euclidean distance, and $\mathcal{Q}$ contains, *in addition* to our $m+n$ plurimers, *their mirror images in the vertical line* touching all zig-zags on the right boundary of W_N, and also an *up-pointing monomer* just outside the lower of the two aligned consecutive segments of this right boundary.

Relation (2.11) also has a strong electrostatic reminiscence. It shows that the quadromers near a zig-zag boundary can be interpreted as electrical charges near a straight line whose effect is to bring in the mirror images of our charges. A well-known situation in which this happens is for electrical charges near a conductor—but in that case the image charges need to be taken of *opposite* sign. In fact, there is another physical situation in which the images are taken with the same sign: that when we have fixed charges inside a dielectric and near a straight boundary separating this dielectric from another one, of a dielectric constant negligible in comparison with the first (see e.g. [11, §12-2]). Our relation (2.11) models this situation—with the additional specification that the slight irregularity of the right boundary near its middle behaves like an extra charge of magnitude +1.

We conclude this section by presenting a view of our results as a possible random tiling model for two dimensional electrostatics.

Suppose one pictures the two dimensional "universe" being spatially quantized, consisting of a very fine lattice of triangular "quanta of space", each of them responsible for creating a unit charge (of sign given by its orientation) when displaced by a "body" (which, since we are considering space to be quantized, can be placed only in such a way that it consists of whole unit triangles). Suppose also that once a number of bodies are placed in this space—thus acquiring the charge determined by the above interpretation—the unoccupied quanta of space (i.e., unit triangles) have the tendency to pair up with their neighbors to form a sea of dimers, according to the uniform distribution on all such possible pairings. (Assuming such a tendency for unoccupied unit triangles to pair up with a neighbor seems natural from the point of view of quantum mechanics, as the quantum fluctuations of the vacuum ceaselessly cause particle-antimatter-companion pairs to erupt into existence only to be annihilated after an instant. One particular rhombus tiling corresponds to one particular way for these virtual particles to annihilate. To consider averaging over all such possible tilings is analogous to Feynman's "sum-over-paths" approach to quantum mechanics; see the remark at the end of Section 15.) Then (2.7) shows—under the assumption that the charges are symmetrically distributed and are all of magnitude ± 2, except for a single $+1$ on the symmetry axis—that these charged bodies would interact precisely like charged bodies in two dimensional electrostatics.

We would get a *proof* of the Superposition Principle (for such charge distributions), and very strong evidence for Coulomb's law![4]

To make this parallelism more explicit, let $\mathcal{S}$ be the family of all $(2m + 2n + 1)$-element sets of triangular plurimers (one monomer, and the rest quadromers) satisfying the hypothesis of Theorem 2.1, and being contained in some fixed disk D centered at the center of the regions (2.1) (the radius of D is some large absolute constant, the "diameter" of our system of plurimers).

Let $\mathcal{A} = \{A_1, \ldots, A_{2m+2n+1}\}$ and $\mathcal{B} = \{B_1, \ldots, B_{2m+2n+1}\}$ be two members of $\mathcal{S}$. Re-denote the charges of their elements by $\mathrm{ch}(A_i) = \mathrm{ch}(B_i) = q_i$, $i = 1, \ldots, 2m + 2n + 1$, and denote their corresponding regions (2.1) by $H_N(\mathcal{A})$ and $H_N(\mathcal{B})$, respectively. Denote also their corresponding correlations (2.2) by $\omega(\mathcal{A})$ and $\omega(\mathcal{B})$, respectively.

Define a probability distribution on $\mathcal{S}$ by requiring the ratio of the probabilities $P_{\mathcal{A}}$ and $P_{\mathcal{B}}$ to be

$$(2.12) \qquad \frac{P_{\mathcal{A}}}{P_{\mathcal{B}}} := \lim_{N \to \infty} \frac{\mathrm{M}(H_N(\mathcal{A}))}{\mathrm{M}(H_N(\mathcal{B}))} = \frac{\omega(\mathcal{A})}{\omega(\mathcal{B})},$$

the second equality being valid by (2.2).

By (2.6) we obtain from (2.12) that

$$\frac{P_{\mathcal{A}}}{P_{\mathcal{B}}} \sim \exp\left(-\frac{1}{2}\left[\sum_{1 \leq i < j \leq 2m+2n+1} q_i q_j (-\ln \mathrm{d}(A_i, A_j)) \right.\right.$$

$$(2.13)$$

$$\left.\left. - \sum_{1 \leq i < j \leq 2m+2n+1} q_i q_j (-\ln \mathrm{d}(B_i, B_j)) \right]\right),$$

for large mutual distances between the elements of $\mathcal{A}$ and $\mathcal{B}$.

Consider, on the other hand, a two dimensional physical system of $2m + 2n + 1$ charges $Q_1, \ldots, Q_{2m+2n+1}$ of magnitudes $q_1 q_e, \ldots, q_{2m+2n+1} q_e$ (expressed as integer multiples of the elementary charge q_e). Then assuming that there are only electrical forces between them (governed by the two dimensional electrostatic potential, $-1/(2\pi\epsilon_0) \ln(d)$, where d is the distance and ϵ_0 is the permittivity of empty space [11, §4-2, §5-5]), we obtain by the Fundamental Theorem of Statistical Physics (see e.g. [10, §40-3]) that the probability $P_e(\mathbf{d})$ of finding the charges at mutual distances $\mathbf{d} = (d_{ij})_{1 \leq i < j \leq 2m+2n+1}$, relative to the probability $P_e(\mathbf{d}')$ of finding them

---

[4]The reason we don't obtain a proof for Coulomb's law from Theorem 2.1 is that we are assuming there a symmetric charge distribution, and therefore the case of two charged bodies in general position is not covered. This case is addressed in [5] for opposite sign charges of magnitude 2. The general even-magnitude case is presented in [6], a sequel of the present paper.

at mutual distances $\mathbf{d}' = (d'_{ij})_{1 \leq i < j \leq 2m+2n+1}$, is

$$\frac{P_e(\mathbf{d})}{P_e(\mathbf{d}')} = \exp\left(-\frac{q_e^2}{2\pi\epsilon_0 kT}\left[\sum_{1 \leq i < j \leq 2m+2n+1} q_i q_j(-\ln d_{ij})\right.\right.$$

(2.14)

$$\left.\left. - \sum_{1 \leq i < j \leq 2m+2n+1} q_i q_j(-\ln d'_{ij})\right]\right),$$

where k is Boltzmann's constant (see e.g. [10, §39-4]) and T is absolute temperature.

Relations (2.13) and (2.14) show that, at least in the case when the two dimensional physical system of electrical charges satisfies the magnitude and geometrical distribution requirements in the hypothesis of Theorem 2.1, their physical electrostatic interaction at temperature $T = q_e^2/(\pi\epsilon_0 k)$ is correctly modeled by our model[5]. We present in Section 15 a possible three dimensional analog of this parallelism that allows any temperature T.

Moreover, what plays the role of *electrostatic potential* in our model is just the *entropy*—the logarithm of the number of dimer coverings—which had been considered before, in a different context [2][23] (it involved the number of dimer coverings of the *inside* structure of a molecule, not of the *outside*, as in our situation), as a possible measure for chemical energy (one of the original motivations was simply its satisfying the addition principle, for isolated systems).

Our interpretation has the advantage that Coulomb's law emerges as an asymptotic law valid when the distances between the charges are large in comparison with the size of the charges. This would explain why Coulomb's law seems to break down at very small distances (cf. [11, §5-8], at distances lower than about 10^{-14}).

Furthermore, in this context (2.11) can be interpreted as describing the interaction of electrical charges with an "edge" of this two dimensional "universe." The same sign for the image charges ensures a repelling effect, and this helps keeping the charges inside the "universe"!

An appealing feature of such a model is that it is discrete, and therefore models the physical electrostatic interaction by considering a discrete ambient space.

---

[5]In fact, if one accepts Conjecture 14.1, our set-up allows one to introduce an extra parameter in (2.13), a positive integer x, which makes the parallelism between (2.13) and (2.14) go through for any temperature. To this end, refine our triangular lattice $\mathcal{T}$ so that each unit triangle is subdivided into equilateral triangles of side $1/x$; denote the new lattice by $\mathcal{T}_x$. The hypotheses of Conjecture 14.1 are clearly satisfied when the plurimers in $\mathcal{A}$ and $\mathcal{B}$ are regarded as plurimers on $\mathcal{T}_x$. Both their charges and their mutual distances get multiplied by x. By (2.13) one readily sees that the effects of the change in distance cancel out, while the effect of changing the charges is that the fraction $-1/2$ in front of the square brackets in (2.13) changes to $-x^2/2$.

A second "calibration" parameter a, a given positive integer, can be introduced as follows. Arrange, by a-fold lattice refinement, for all q_i's in (2.13) to be divisible by a. Factor out the common factor of a^2 in front of the square parenthesis in (2.13) and re-denote each leftover q_i/a by q_i. The formula obtained this way from (2.13) still parallels (2.14), but now the elementary physical charge q_e corresponds to a plurimer of charge a. In particular, a should have a fixed value.

The overall effect of x-fold lattice refinement and "calibration" by a is that the fraction $-1/2$ in front of the square brackets in (2.13) changes to $-(ax)^2/2$.

The possibility of a discrete "machinery" underlying the electrostatics of the real physical world is mentioned by Feynman in [**11**, §12-7]. When indicating how several equations in physics, like for instance that for neutron diffusion, are true only as approximations when the distance over which one looks is large (for neutron diffusion, large in comparison to the mean free path), Feynman goes on to ask:

"Is the same statement perhaps also true for the *electrostatic* equations? Are they also correct only as a smoothed-out imitation of a really much more complicated microscopic world? Could it be that the real world consists of little X-ons which can be seen only at *very* tiny distances? And that in our measurements we are always observing on such a large scale that we can't see these little X-ons, and that is why we get the differential equations?

Our currently most complete theory of electrodynamics does indeed have its difficulties at very short distances. So it is possible, in principle, that these equations are smoothed-out versions of something. They appear to be correct at distances down to about 10^{-14} cm, but then they begin to look wrong. It is possible that there is some as yet undiscovered underlying 'machinery,' and that the details of an underlying complexity are hidden in the smooth-looking equations—as is so in the 'smooth' diffusion of neutrons. But no one has yet formulated a successful theory that works that way."

The random tiling model presented in this paper (and pursued further in a subsequent paper) seems to be a possible such "machinery" that would produce, in a two dimensional world, precisely the effects of electrostatics. As presented in Section 14, we conjecture that the validity of our model does not depend on the particular choice of the hexagonal lattice, but it holds in fact for any plane bipartite lattice under a suitable embedding. This would imply that the electrostatic effects we obtain depend only on the *space*—in accordance with Feynman's suggestion [**11**, §12-7] that it might be the space itself, the common "framework into which physics is put," that is responsible for the emergence of such a simple equation governing electrostatics (as well as other physical phenomena).

The model presented in this paper has natural analogs in higher dimensions—for instance, by considering the cubic lattices. We believe that the three dimensional analog presents the effects of electrostatics in the real three dimensional world. Some details about these considerations are presented in Section 15.

Reduction to boundary-influenced correlations

We will find it convenient to express the number of dimer coverings of the region $H_N \begin{pmatrix} R_1 & \dots & R_m \\ v_1 & & v_m \end{pmatrix}; \begin{matrix} R_1' & \dots & R_n' \\ v_1' & & v_n' \end{matrix}$ in terms of the number of tilings of two "halves" of it. One of them is the region $W_N \begin{pmatrix} R_1 & \dots & R_m \\ v_1 & & v_m \end{pmatrix}; \begin{matrix} R_1' & \dots & R_n' \\ v_1' & & v_n' \end{matrix}$ defined in Section 2— the portion to the west of the cut illustrated in Figure 2.3. The other is the part of H_N to the east of that cut—more precisely, what is left from the region east of the cut after removing the forced dimers (see Figure 3.1 for an example; the forced dimers are shaded), and weighting by $1/2$ the $N + 2m - 1$ dimer positions below the monomer u that are closest to the cut; denote it by $E_N \begin{pmatrix} R_1 & \dots & R_m \\ v_1 & & v_m \end{pmatrix}; \begin{matrix} R_1' & \dots & R_n' \\ v_1' & & v_n' \end{matrix}$.

Figure 3.1 illustrates an example (for clarity, the two regions are pictured after separating them horizontally by one unit; the dimer positions weighted by $1/2$ are indicated by shaded ellipses).

The regions W_N and E_N were defined in such a way that they are precisely the ones that result from applying the Factorization Theorem [3, Theorem 1.2] to the region H_N. Therefore, the quoted Factorization Theorem yields the following result.

PROPOSITION 3.1. *We have*

$$\mathrm{M}\left(H_N \begin{pmatrix} R_1 & \dots & R_m \\ v_1 & & v_m \end{pmatrix}; \begin{matrix} R_1' & \dots & R_n' \\ v_1' & & v_n' \end{matrix} \right) = 2^{N+m+n}$$

$$\times \mathrm{M}\left(W_N \begin{pmatrix} R_1 & \dots & R_m \\ v_1 & & v_m \end{pmatrix}; \begin{matrix} R_1' & \dots & R_n' \\ v_1' & & v_n' \end{matrix} \right) \mathrm{M}\left(E_N \begin{pmatrix} R_1 & \dots & R_m \\ v_1 & & v_m \end{pmatrix}; \begin{matrix} R_1' & \dots & R_n' \\ v_1' & & v_n' \end{matrix} \right).$$

(3.1)

In particular, (3.1) holds for the reference position of the plurimers used in definitions (2.2) and (2.8). We obtain

$$\mathrm{M}\left(H_N \begin{pmatrix} 1 & \dots & 2m-1 \\ 0 & & 0 \end{pmatrix}; \begin{matrix} 1 & \dots & 2n-1 \\ 0 & & 0 \end{matrix} \right) = 2^{N+m+n}$$

$$\times \mathrm{M}\left(W_N \begin{pmatrix} 1 & \dots & 2m-1 \\ 0 & & 0 \end{pmatrix}; \begin{matrix} 1 & \dots & 2n-1 \\ 0 & & 0 \end{matrix} \right) \mathrm{M}\left(E_N \begin{pmatrix} 1 & \dots & 2m-1 \\ 0 & & 0 \end{pmatrix}; \begin{matrix} 1 & \dots & 2n-1 \\ 0 & & 0 \end{matrix} \right).$$

(3.2)

Dividing (3.1) and (3.2) side by side and letting $N \to \infty$, one sees by (2.2) and (2.8) that ω and ω_b can be related by considering another boundary-influenced

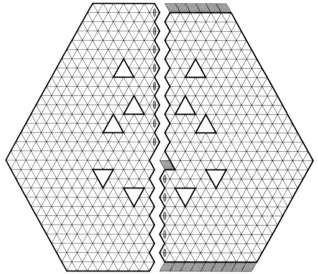

FIGURE 3.1. $W_2 \begin{pmatrix} 5 & 2 & 4 & 2 & 3 \\ 0 & 1 & 1 & 2 & 4 \end{pmatrix}$ and $E_2 \begin{pmatrix} 5 & 2 & 4 & 2 & 3 \\ 0 & 1 & 1 & 2 & 4 \end{pmatrix}$.

correlation, defined by means of the regions E_N:

(3.3)

$$
\bar{\omega}_b \begin{pmatrix} R_1 & \cdots & R_m & R_1' & \cdots & R_n' \\ v_1 & & v_m & v_1' & & v_n' \end{pmatrix} := \lim_{N \to \infty} \frac{\mathrm{M}\left(E_N \begin{pmatrix} R_1 & \cdots & R_m & R_1' & \cdots & R_n' \\ v_1 & & v_m & v_1' & & v_n' \end{pmatrix} \right)}{\mathrm{M}\left(E_N \begin{pmatrix} 1 & \cdots & 2m-1 & 1 & \cdots & 2n-1 \\ 0 & & 0 & 0 & & 0 \end{pmatrix} \right)}
$$

(the fact that this limit exists follows by Lemma 13.1).

By (3.1)–(3.3), the definitions (2.2) and (2.8), and the existence of the limits in the latter two definitions (guaranteed by Lemma 5.1 and Lemma 13.1), one obtains the following result.

PROPOSITION 3.2. *The limit* (2.2) *defining the correlation at the center exists and its value is the product of the two boundary-influenced correlations:*

$$
\omega \begin{pmatrix} R_1 & \cdots & R_m & R_1' & \cdots & R_n' \\ v_1 & & v_m & v_1' & & v_n' \end{pmatrix} =
$$

(3.4) $\qquad \omega_b \begin{pmatrix} R_1 & \cdots & R_m & R_1' & \cdots & R_n' \\ v_1 & & v_m & v_1' & & v_n' \end{pmatrix} \bar{\omega}_b \begin{pmatrix} R_1 & \cdots & R_m & R_1' & \cdots & R_n' \\ v_1 & & v_m & v_1' & & v_n' \end{pmatrix}.$

Therefore, in order to obtain the asymptotic behavior of the correlation at the center ω, it is enough to study the boundary-influenced correlations ω_b and $\bar{\omega}_b$.

4

A simple product formula for correlations along the boundary

Our calculations are built upon an explicit product formula that we present in this section for certain correlations along the boundary of the regions W_N. To state this we need to introduce a new family of regions, closely related to the W_N's, and in terms of which the boundary-influenced correlation ω_b turns out to be expressible.

Let W be the region determined by the common outside boundary of the regions $W_N\begin{pmatrix} R_1 & \cdots & R_m & R'_1 & \cdots & R'_n \\ v_1 & & v_m & v'_1 & & v'_n \end{pmatrix}$, for fixed N, m and n. Then W is the half-hexagonal lattice region with four straight sides—the southern side of length $N + 2n$, southwestern of length $2N + 4m$, northwestern of length $2N + 4n + 1$, and northern of length $N + 2m$—followed by $N + 2n$ descending zig-zags to the lattice point O, one extra unit step southwest of O, and $N + 2m$ more descending zig-zags to close up the boundary (the boundary of the region W corresponding to $m = 4$, $n = 6$, and $N = 2$ can be seen in Figure 4.1). In addition, the $N + 2n$ dimer positions weighted $1/2$ in the regions W_N are also weighted so in W.

The eastern side of W can be viewed as consisting of *bumps*—pairs of adjacent lattice segments forming an angle that opens to the west: $N + 2m$ bumps below O, and $N + 2n$ above O. Label the former by $0, 1, \ldots, N + 2m - 1$ and the latter by $0, 1, \ldots, N + 2n - 1$, both labelings starting with the bumps closest to O and then moving successively outwards.

Since everywhere in the description of W the parameters m and n appear with even multiplicative coefficients, we re-denote, for notational simplicity, $2m$ by m and $2n$ by n. Therefore we consider the four straight sides of W to have lengths $N + n$, $2N + 2m$, $2N + 2n + 1$ and $N + m$, while the number of bumps below and above O is $N + m$ and $N + n$, respectively. The results of this section do not assume that m and n are even (even though we will only use them for even m and n).

We allow any bump above O to be "removed" by placing an up-pointing quadromer (lattice triangle of side two) across it and discarding the three monomers of W it covers. Similarly, a bump below O can be removed by placing a down-pointing quadromer across it and discarding the three monomers it covers.

We are now ready to introduce the family of regions mentioned in the first paragraph of this section: define $W_N[k_1, \ldots, k_m; l_1, \ldots, l_n]$ to be the region obtained from W by removing the bumps below O with labels $0 \leq k_1 < k_2 < \cdots < k_m \leq N + m - 1$, and the bumps above O with labels $0 \leq l_1 < l_2 < \cdots < l_n \leq N + n - 1$. Figure 4.1 shows $W_2[1, 2, 3, 5; 0, 1, 3, 4, 5, 6]$.

The product formula referred to in the title of this section is stated in the following result.

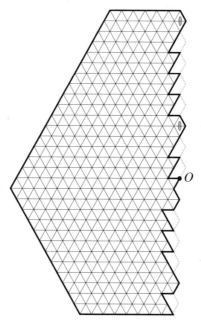

FIGURE 4.1. $W_2[1,2,3,5;0,1,3,4,5,6]$.

PROPOSITION 4.1. *For $m,n \geq 0$ and fixed integers $0 \leq k_1 < k_2 < \cdots < k_m$ and $0 \leq l_1 < l_2 < \cdots < l_n$ we have*

$$\lim_{N\to\infty} \frac{\mathrm{M}\left(W_N[k_1,\ldots,k_m;l_1,\ldots,l_n]\right)}{\mathrm{M}\left(W_N[0,\ldots,m-1;0,\ldots,n-1]\right)} =$$

(4.1)
$$\prod_{i=1}^{m} \frac{\frac{(3/2)_{k_i}}{(2)_{k_i}}}{\frac{(3/2)_{i-1}}{(2)_{i-1}}} \prod_{i=1}^{n} \frac{\frac{(3/2)_{l_i}}{(1)_{l_i}}}{\frac{(3/2)_{i-1}}{(1)_{i-1}}} \frac{\displaystyle\prod_{1\leq i<j\leq m} \frac{k_j-k_i}{j-i} \prod_{1\leq i<j\leq n} \frac{l_j-l_i}{j-i}}{\displaystyle\prod_{i=1}^{m}\prod_{i=1}^{n} \frac{k_i+l_j+2}{i+j}}$$

(4.2)
$$= \chi_{m,n} \prod_{i=1}^{m} \frac{(3/2)_{k_i}}{(2)_{k_i}} \prod_{i=1}^{n} \frac{(3/2)_{l_i}}{(1)_{l_i}} \frac{\displaystyle\prod_{1\leq i<j\leq m}(k_j-k_i) \prod_{1\leq i<j\leq n}(l_j-l_i)}{\displaystyle\prod_{i=1}^{m}\prod_{i=1}^{n}(k_i+l_j+2)},$$

where $(a)_k := a(a+1)\cdots(a+k-1)$ (by convention, $(a)_0 := 1$) and

(4.3)
$$\chi_{m,n} = \prod_{i=0}^{m-1} \frac{(2)_i}{(1)_i(3/2)_i} \prod_{i=0}^{n-1} \frac{(i+2)_m}{(3/2)_i}.$$

PROOF. An explicit product formula for $\mathrm{M}\left(W_N[k_1,\ldots,k_m;l_1,\ldots,l_n]\right)$ follows from the results of the paper in Part B of this Memoir. Indeed, our regions $W_N[k_1,\ldots,k_m;l_1,\ldots,l_n]$ are just a different notation for the regions $\bar{R}_{\mathbf{k},\mathbf{l}}(x)$ defined in Section 2 of that paper: In the notation from there, x is the length of the base—$N+n$ in the current notation—while $\mathbf{k}$ and $\mathbf{l}$ are the (strictly increasing) lists of labels—incremented by 1, due to a shift of one unit in labeling bumps in the present paper compared to the paper in Part B—of the bumps that *remain* in the region. Therefore,

$$W_N[k_1,\ldots,k_m;l_1,\ldots,l_n] =$$
(4.4)
$$\bar{R}_{[1,\ldots,N+m]\backslash[k_1+1,\ldots,k_m+1],[1,\ldots,N+n]\backslash[l_1+1,\ldots,l_n+1]}(N+n).$$

Relations (6.1), (6.2) and (6.4) from the paper in Part B of this Memoir state that for any pair of lists $\mathbf{p} = [p_1,\ldots,p_s]$, $1 \leq p_1 < p_2 < \ldots < p_s$ and $\mathbf{q} = [q_1,\ldots,q_t]$, $1 \leq q_1 < q_2 < \ldots < q_t$, and for any nonnegative integer $x \geq q_t - p_s - t + s$ (see Section 2 of Part B), one has

(4.5)
$$\mathrm{M}\left(\bar{R}_{\mathbf{p},\mathbf{q}}(x)\right) = \bar{c}_{\mathbf{p},\mathbf{q}}F_{\mathbf{p},\mathbf{q}}(x),$$

where

(4.6) $$\bar{c}_{\mathbf{p},\mathbf{q}} = 2^{\binom{t-s}{2}-s}\prod_{i=1}^{s}\frac{1}{(2p_i-1)!}\prod_{i=1}^{t}\frac{1}{(2q_i)!}\frac{\prod_{1\leq i<j\leq s}(p_j-p_i)\prod_{1\leq i<j\leq t}(q_j-q_i)}{\prod_{i=1}^{s}\prod_{j=1}^{t}(p_i+q_j)}$$

and the polynomials $F_{\mathbf{p},\mathbf{q}}(x)$ satisfy

(4.7)
$$\frac{F_{\mathbf{p}^{|i\rangle},\mathbf{q}}(x)}{F_{\mathbf{p},\mathbf{q}}(x)} = (x-p_i+p_s)(x+p_i+p_s-s+t+1), \quad \text{for } 1 \leq i < s$$

(4.8)
$$\frac{F_{\mathbf{p},\mathbf{q}^{|i\rangle}}(x)}{F_{\mathbf{p},\mathbf{q}}(x)} = (x+q_i+p_s+1)(x-q_i+p_s-s+t), \quad \text{for } 1 \leq i \leq t$$

(here $\mathbf{p}^{|i\rangle}$ is the list obtained from $\mathbf{p}$ by increasing its i-th element by 1—in particular, it is defined only if $l_{i+1} - l_i \geq 2$).

To deduce the limit (4.1) from these formulas, it will be convenient to consider first the limit

(4.9)
$$\lim_{N\to\infty}\frac{\mathrm{M}\left(W_N[k_1,\ldots,k_{i-1},k_i+1,k_{i+1},\ldots,k_m;l_1,\ldots,l_n]\right)}{\mathrm{M}\left(W_N[k_1,\ldots,k_{i-1},k_i,k_{i+1},\ldots,k_m;l_1,\ldots,l_n]\right)},$$

for $k_i + 1 < k_{i+1}$.

Use (4.4) to view the regions involved in this fraction as $\bar{R}_{\mathbf{p},\mathbf{q}}(x)$'s. The lists $\mathbf{p}$ and $\mathbf{q}$ corresponding to the regions at the numerator and denominator in (4.9) are illustrated in Figure 4.2 (the shaded squares indicate the entries removed at the indices on the right hand side of (4.4)). Since the $\mathbf{q}$-lists are the same, the limit (4.9) can be found applying formulas (4.6) and (4.7) to the lists illustrated in Figure 4.2.

Since the only difference between the $\mathbf{p}$-lists in Figure 4.2(a) is the one indicated by the dot in that figure, we obtain from (4.6) that the contribution to the fraction

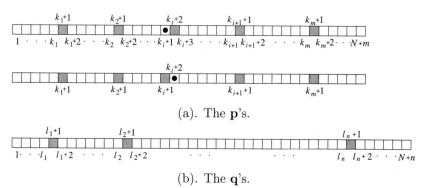

(a). The **p**'s.

(b). The **q**'s.

FIGURE 4.2. Incremental change in the first index list.

in (4.9) coming from the ratio of the $\bar{c}_{\mathbf{p},\mathbf{q}}$'s is

$$\frac{\dfrac{1}{(2k_i+1)!}}{\dfrac{1}{(2k_i+3)!}} \frac{\dfrac{k_i!}{(k_i-k_1)\cdots(k_i-k_{i-1})}}{\dfrac{(k_i+1)!}{(k_i-k_1+1)\cdots(k_i-k_{i-1}+1)}} \frac{\dfrac{[N+m-(k_i+1)]!}{(k_{i+1}-k_i)\cdots(k_m-k_i)}}{\dfrac{[N+m-(k_i+2)]!}{(k_{i+1}-k_i-1)\cdots(k_m-k_i-1)}}$$

$$\times \frac{\dfrac{(k_i+3)\cdots(k_i+2+N+m)}{(k_i+l_1+3)\cdots(k_i+l_n+3)}}{\dfrac{(k_i+2)\cdots(k_i+1+N+m)}{(k_i+l_1+2)\cdots(k_i+l_n+2)}}$$

$$= (k_i+2+N+n)(N+m-(k_i+1))\frac{(2k_i+2)(2k_i+3)}{(k_i+1)(k_i+2)}$$

$$\times \frac{((k_i+1)-k_1)\cdots((k_i+1)-k_{i-1})(k_{i+1}-(k_i+1))\cdots(k_m-(k_i+1))}{(k_i-k_1)\cdots(k_i-k_{i-1})(k_{i+1}-k_i)\cdots(k_m-k_i)}$$

(4.10)

$$\times \frac{(k_i+l_1+2)\cdots(k_i+l_n+2)}{(k_i+l_1+3)\cdots(k_i+l_n+3)}.$$

On the other hand, the contribution to the fraction in (4.9) coming from the polynomials $F_{\mathbf{p},\mathbf{q}}(x)$ is readily seen from (4.7) to be

(4.11)
$$\frac{1}{(2N+m+n-(k_i+1))(2N+m+n+(k_i+1)+1)}.$$

By (4.10) and (4.11) one readily finds the limit (4.9) to be

$$\lim_{N\to\infty} \frac{\mathrm{M}\left(W_N[k_1,\ldots,k_{i-1},k_i+1,k_{i+1},\ldots,k_m;l_1,\ldots,l_n]\right)}{\mathrm{M}\left(W_N[k_1,\ldots,k_{i-1},k_i,k_{i+1},\ldots,k_m;l_1,\ldots,l_n]\right)} =$$

$$\frac{k_i+3/2}{k_i+2} \frac{((k_i+1)-k_1)\cdots((k_i+1)-k_{i-1})(k_{i+1}-(k_i+1))\cdots(k_m-(k_i+1))}{(k_i-k_1)\cdots(k_i-k_{i-1})(k_{i+1}-k_i)\cdots(k_m-k_i)}$$

(4.12)

$$\times \frac{(k_i+l_1+2)\cdots(k_i+l_n+2)}{((k_i+1)+l_1+2)\cdots((k_i+1)+l_n+2)}.$$

Regard (4.12) as giving the effect of decrementing the entry $k_i + 1$ in the argument to k_i. A further decrementation of this entry to $k_i - 1$ will produce, by (4.12), the factor

$$\frac{k_i + 1/2}{k_i + 1} \frac{(k_i - k_1)\cdots(k_i - k_{i-1})(k_{i+1} - k_i)\cdots(k_m - k_i))}{((k_i - 1) - k_1)\cdots((k_i - 1) - k_{i-1})(k_{i+1} - (k_i - 1))\cdots(k_m - (k_i - 1))}$$

(4.13)

$$\times \frac{(k_i + l_1 + 1)\cdots(k_i + l_n + 1)}{(k_i + l_1 + 2)\cdots(k_i + l_n + 2)}.$$

One notices that telescoping simplifications occur when multiplying together the second and third fractions in (4.12) for successive decrementations of each argument k_i. In particular, if the argument k_1 is decremented all the way to 0, we obtain by repeated application of (4.12) that

$$\lim_{N \to \infty} \frac{M(W_N[k_1, k_2, \ldots, k_m; l_1, \ldots, l_n])}{M(W_N[0, k_2, \ldots, k_m; l_1, \ldots, l_n])} =$$

(4.14)

$$\frac{(3/2)_{k_1}}{(2)_{k_1}} \frac{(k_2 - k_1)(k_3 - k_1)\cdots(k_m - k_1)}{(k_2 - 0)(k_3 - 0)\cdots(k_m - 0)} \frac{(l_1 + 2)\cdots(l_n + 2)}{(k_1 + l_1 + 2)\cdots(k_1 + l_n + 2)}.$$

Similarly, we obtain

$$\lim_{N \to \infty} \frac{M(W_N[0, k_2, k_3, \ldots, k_m; l_1, \ldots, l_n])}{M(W_N[0, 1, k_3, \ldots, k_m; l_1, \ldots, l_n])} =$$

$$\frac{(5/2)_{k_2-1}}{(3)_{k_2-1}} \frac{(k_2 - 0)(k_3 - k_2)\cdots(k_m - k_2)}{(1 - 0)(k_3 - 1)\cdots(k_m - 1)} \frac{(l_1 + 3)\cdots(l_n + 3)}{(k_2 + l_1 + 2)\cdots(k_2 + l_n + 2)},$$

$$\lim_{N \to \infty} \frac{M(W_N[0, 1, k_3, k_4, \ldots, k_m; l_1, \ldots, l_n])}{M(W_N[0, 1, 2, k_4, \ldots, k_m; l_1, \ldots, l_n])} =$$

$$\frac{(7/2)_{k_3-2}}{(4)_{k_3-2}} \frac{(k_3 - 0)(k_3 - 1)(k_4 - k_3)\cdots(k_m - k_3)}{(2 - 0)(2 - 1)(k_4 - 2)\cdots(k_m - 2)} \frac{(l_1 + 4)\cdots(l_n + 4)}{(k_3 + l_1 + 2)\cdots(k_3 + l_n + 2)},$$

$$\vdots$$

$$\lim_{N \to \infty} \frac{M(W_N[0, 1, \ldots, m - 2, k_m; l_1, \ldots, l_n])}{M(W_N[0, 1, \ldots, m - 1; l_1, \ldots, l_n])} =$$

$$\frac{((2m+1)/2)_{k_m-m+1}}{(m+1)_{k_m-m+1}} \frac{(k_m - 0)(k_m - 1)\cdots(k_m - (m - 2))}{((m - 1) - 0)((m - 1) - 1)\cdots((m - 1) - (m - 2))}$$

(4.15)

$$\times \frac{(l_1 + m + 1)\cdots(l_n + m + 1)}{(k_m + l_1 + 2)\cdots(k_m + l_n + 2)}.$$

Now multiply together the equalities in (4.14) and (4.15). The first fractions on their right hand sides combine to give $\prod_{i=1}^{m}((3/2)_{k_i}/(2)_{k_i})/\prod_{i=1}^{m}((3/2)_{i-1}/(2)_{i-1})$. Due

k_1+1 $\qquad$ k_2+1 $\qquad\qquad$ k_m+1

$1\cdot\ \cdot k_1\ \ k_1+2\ \ \cdots\ \ k_2\ \ k_2+2\qquad\cdots\qquad k_m\ \ k_m+2\ \cdots\ N+m$

(a). The **p**'s.

l_1+1 $\qquad$ l_1+1 $\qquad$ l_i+2 $\quad l_{i+1}+1$ $\qquad l_m+1$

$1\ \cdots\ l_1\ \ l_1+2\cdots\ \ l_2\ \ l_2+2\cdot\ \cdot\ \ l_i+1\ \ l_i+3\ \ \cdots\ \ l_{i+1}\ l_{i+1}+2\ \cdots\ l_m\ \ l_m+2\ \ \cdots N+n$

l_i+2

$l_1+1\qquad\quad l_2+1\qquad\quad l_i+1\qquad\quad l_{i+1}+1\qquad\quad l_n+1$

(b). The **q**'s.

FIGURE 4.3. Incremental change in the second index list.

to simplifications, the second fractions on the right hand sides yield $\prod_{1\leq i<j\leq m}(k_j-k_i)/(j-i)$. The third fractions give $\prod_{i=1}^{n}(l_i+2)m/\prod_{i=1}^{m}\prod_{j=1}^{n}(k_i+l_j+2)$. We obtain

$$\lim_{N\to\infty}\frac{\mathrm{M}\left(W_N[k_1,k_2,\ldots,k_m;l_1,\ldots,l_n]\right)}{\mathrm{M}\left(W_N[0,1,\ldots,m-1;l_1,\ldots,l_n]\right)}=\prod_{i=1}^{n}(l_i+2)_m$$

(4.16)
$$\times\ \frac{\prod_{i=1}^{m}\dfrac{(3/2)_{k_i}}{(2)_{k_i}}}{\prod_{i=1}^{m}\dfrac{(3/2)_{i-1}}{(2)_{i-1}}}\ \frac{\prod_{1\leq i<j\leq m}\dfrac{k_j-k_i}{j-i}}{\prod_{i=1}^{m}\prod_{j=1}^{n}(k_i+l_j+2)}.$$

The effect of decrementing arguments in the list $[l_1,\ldots,l_n]$ can be analyzed in a similar way. Indeed, compare the situation when the lists of k_j's at the indices on the left hand side of (4.1) are the same, and the lists of l_j's are $[l_1,\ldots,l_{i-1},l_i+1,l_{i+1},\ldots,l_n]$ and $[l_1,\ldots,l_{i-1},l_i,l_{i+1},\ldots,l_n]$, respectively. These lists are illustrated in Figure 4.3. We need to find

(4.17)
$$\lim_{N\to\infty}\frac{\mathrm{M}\left(W_N[k_1,\ldots,k_m;l_1,\ldots,\ldots,l_{i-1},l_i+1,l_{i+1},l_n]\right)}{\mathrm{M}\left(W_N[k_1,\ldots,k_m;l_1,\ldots,l_{i-1},l_i,l_{i+1},\ldots,l_n]\right)},$$

for $l_i+1<l_{i+1}$.

By (4.4), view the regions in (4.17) as $\bar{R}_{\mathbf{p},\mathbf{q}}(x)$'s and use formulas (4.5), (4.6) and (4.8). Repeating the reasoning that proved (4.12), we see that the formulas we obtain now are almost exactly (4.10) and (4.11), with the roles of the lists $[k_1,\ldots,k_m]$ and $[l_1,\ldots,l_n]$ interchanged.

Indeed, the only difference from (4.10) of the contribution coming from the $\bar{c}_{\mathbf{p},\mathbf{q}}$'s of (4.6) is that the first fraction on the left hand side of the analog of (4.10) is $(1/(2(l_i+1))!)/(1/(2(l_i+2))!)$, as opposed to the $(1/(2l_i+1)!)/(1/(2l_i+3)!)$ that results from (4.10) by interchanging the lists $[k_1,\ldots,k_m]$ and $[l_1,\ldots,l_n]$. The consequence of this difference is that the numerator of the first fraction after the equality sign in the present situation analog of (4.10) is $(2l_i+3)(2l_i+4)$, and therefore the first fraction on the right hand side of the analog of (4.12) is $(l_i+3/2)/(l_i+1)$.

On the other hand, the contribution to the left hand side of (4.17) coming from the $F_{\mathbf{p},\mathbf{q}}$'s of (4.6) is readily seen to have, after letting $N\to\infty$, *exactly* the same effect as (4.11).

Therefore formulas (4.14) and (4.15) have perfect analogs when changing the arguments of the second list, with the only difference that the integers between the round parentheses at the denominators in the first fractions on their right hand sides are now decremented by one unit: they are $(1)_{l_1}, (2)_{l_2-1}, \ldots, (n)_{l_n-n+1}$.

Just as we deduced (4.16) from (4.14) and (4.15), we obtain from the above analysis of the differences between decrementing the k_i's and the l_j's that

$$\lim_{N \to \infty} \frac{\mathrm{M}\left(W_N[k_1, \ldots, k_m; l_1, l_2, \ldots, l_n]\right)}{\mathrm{M}\left(W_N[k_1, \ldots, k_m; 0, 1, \ldots, n-1]\right)} = \prod_{i=1}^{m} (k_i + 2)_n$$

(4.18)
$$\times \frac{\prod_{i=1}^{n} \frac{(3/2)_{l_i}}{(1)_{l_i}}}{\prod_{i=1}^{n} \frac{(3/2)_{i-1}}{(1)_{i-1}}} \frac{\prod_{1 \le i < j \le n} \frac{l_j - l_i}{j - i}}{\prod_{i=1}^{m} \prod_{j=1}^{n} (k_i + l_j + 2)}.$$

Combining (4.16) with the specialization of (4.18) for $k_i = i - 1$, $i = 1, \ldots, m$, and using $\prod_{i=1}^{m}((i-1)+2)_n = \prod_{i=1}^{m} \prod_{j=1}^{n}(i+j)$, we obtain the first equality in (4.1). The second equality is just a repackaging of the first, using $\prod_{1 \le i < j \le m}(j - i) = \prod_{i=1}^{m}(1)_{i-1}$ and $\prod_{i=1}^{m} \prod_{j=1}^{n}(i+j) = \prod_{j=0}^{n-1}(j+2)_n$. $\qquad\square$

A $(2m + 2n)$-fold sum for ω_b

In this section we present an expression for the boundary-influenced correlation ω_b as a $(2m+2n)$-fold sum. We deduce this by expressing the generating function for dimer coverings of the region $W_N \begin{pmatrix} R_1 & \dots & R_m \, ; & R'_1 & \dots & R'_n \\ v_1 & & v_m & ; & v'_1 & & v'_n \end{pmatrix}$ in terms of generating functions for the dimer coverings of the regions $W_N[k_1, \dots, k_m; l_1, \dots, l_n]$ introduced in the previous section, and then using the explicit product formula (4.1). The remaining part of the paper will consist mainly of analyzing the asymptotics of this multiple sum.

The result mentioned in the title of this section is the following.

LEMMA 5.1. *For fixed* $R_1, \dots, R_m, R'_1, \dots, R'_n \geq 1$ *and* $v_1, \dots, v_m, v'_1, \dots, v'_n \geq 0$ *we have*

$$
\omega_b \begin{pmatrix} R_1 & \dots & R_m \, ; & R'_1 & \dots & R'_n \\ v_1 & & v_m & ; & v'_1 & & v'_n \end{pmatrix} = \chi_{2m,2n} \prod_{i=1}^{m} R_i \prod_{i=1}^{n} R'_i (R'_i - 1/2)(R'_i + 1/2)
$$

$$
\times \left| \sum_{a_1, b_1 = 0}^{R_1} \dots \sum_{a_m, b_m = 0}^{R_m} \sum_{c_1, d_1 = 0}^{R'_1} \dots \sum_{c_n, d_n = 0}^{R'_n} (-1)^{\sum_{i=1}^{m}(a_i + b_i) + \sum_{i=1}^{n}(c_i + d_i)} \right.
$$

$$
\times \prod_{i=1}^{m} \frac{(R_i + a_i - 1)! \, (R_i + b_i - 1)!}{(2a_i)! \, (R_i - a_i)! \, (2b_i)! \, (R_i - b_i)!} \frac{(3/2)_{v_i + a_i} \, (3/2)_{v_i + b_i}}{(2)_{v_i + a_i} \, (2)_{v_i + b_i}}
$$

$$
\times \prod_{i=1}^{n} \frac{(R_i + c_i - 1)! \, (R_i + d_i - 1)!}{(2c_i + 1)! \, (R_i - c_i)! \, (2d_i + 1)! \, (R_i - d_i)!} \frac{(3/2)_{v'_i + c_i} \, (3/2)_{v'_i + d_i}}{(1)_{v'_i + c_i} \, (1)_{v'_i + d_i}}
$$

$$
\times \prod_{1 \leq i < j \leq m} (v_j - v_i + a_j - a_i)(v_j - v_i + a_j - b_i)(v_j - v_i + b_j - a_i)(v_j - v_i + b_j - b_i)
$$

$$
\times \prod_{1 \leq i < j \leq n} (v'_j - v'_i + c_j - c_i)(v'_j - v'_i + c_j - d_i)(v'_j - v'_i + d_j - c_i)(v'_j - v'_i + d_j - d_i)
$$

$$
\times \left. \frac{\prod_{i=1}^{m}(a_i - b_i)^2 \prod_{i=1}^{n}(c_i - d_i)^2}{\prod_{i=1}^{m} \prod_{j=1}^{n}(u_{ij} + a_i + c_j)(u_{ij} + a_i + d_j)(u_{ij} + b_i + c_j)(u_{ij} + b_i + d_j)} \right|,
$$

(5.1)

where

$$
u_{ij} = v_i + v'_j + 2
$$

for $i = 1, \dots, m$, $j = 1, \dots, n$, *and* $\chi_{m,n}$ *is given by* (4.3).

To prove this Lemma we will need the following special case of the Lindström-Gessel-Viennot theorem on non-intersecting lattice paths (see e.g. [13] or [29]).

Consider lattice paths on the directed grid graph $\mathbb{Z}^2$, with edges oriented so that they point in the positive coordinate directions. We allow the edges of $\mathbb{Z}^2$ to be weighted, and define the weight of a lattice path to be the product of the weights on its steps. The weight of an N-tuple of lattice paths is the product of the individual weights of its members. The weighted count of a set of N-tuples of lattice paths is the sum of the weights of its elements.

Let $\mathbf{u} = (u_1, \ldots, u_N)$ and $\mathbf{v} = (v_1, \ldots, v_N)$ be two fixed sets of starting and ending points on $\mathbb{Z}^2$, and let $\mathcal{N}(\mathbf{u}, \mathbf{v})$ be the set of non-intersecting N-tuples of lattice paths with these starting and ending points. For $\mathbf{P} \in \mathcal{N}(\mathbf{u}, \mathbf{v})$, let $\sigma_{\mathbf{P}}$ be the permutation induced by $\mathbf{P}$ on the set consisting of the N indices of its starting and ending points.

THEOREM 5.2 (LINDSTRÖM-GESSEL-VIENNOT).

$$\sum_{\mathbf{P} \in \mathcal{N}(\mathbf{u}, \mathbf{v})} (-1)^{\sigma_{\mathbf{P}}} \operatorname{wt}(\mathbf{P}) = \det\left((a_{ij})_{1 \le i, j \le n}\right),$$

where a_{ij} is the weighted count of the lattice paths from u_i to v_j.

What makes possible the use of this result in our setting is a well-known procedure of encoding dimer coverings by families of non-intersecting "paths of dimers:" given a dimer covering T of a region R on the triangular lattice and a lattice line direction d, the dimers of T parallel to d (i.e., having two sides parallel to d) can naturally be grouped into non-intersecting paths joining the lattice segments on the boundary of R that are parallel to d, and conversely this family of paths determines the dimer covering (see Figure 5.1 for an illustration of this, and e.g. the paper in Part B of this Memoir for a more detailed account).

Consider such a path of dimers P. Let $\mathcal{T}$ denote our triangular lattice. Clearly, P can be identified with a lattice path on the lattice $\mathcal{L}$ of rhombi formed by the midpoints of the segments of $\mathcal{T}$ that are parallel to the encoding direction d. In turn, by deforming the lattice of rhombi $\mathcal{L}$ to a square lattice, the path of dimers P can be regarded as a lattice path on $\mathbb{Z}^2$. It is in this sense that we will view the paths of dimers as lattice paths on $\mathbb{Z}^2$ in the remainder of this section.

PROOF OF LEMMA 5.1. Choose the lattice line direction d in the above encoding procedure to be the southwest-northeast direction of $\mathcal{T}$. Encode the dimer coverings of $W_N \begin{pmatrix} R_1 & \ldots & R_m & R_1' & \ldots & R_n' \\ v_1 & & v_m & v_1' & & v_n' \end{pmatrix}$ by $(2N + 2m + 4n + 1)$-tuples of non-intersecting paths—$2N + 4n + 1$ starting at the northwestern side, and $2m$ starting at the m down-pointing removed quadromers—consisting of dimers parallel to d (see Figure 5.1; there and in the following figures the $N + 2n$ dimer positions weighted by $1/2$ are not distinguished, but are understood to carry that weight).

Let $\mathbf{P}$ be such a $(2N+2m+4n+1)$-tuple. Consider the permutation $\sigma_{\mathbf{P}}$ induced by $\mathbf{P}$ on the set of the $2N + 2m + 4n + 1$ indices of its starting and ending points. It is easy to see that the fact that each quadromer contributes either two starting points on consecutive lattice segments or two analogous ending points, together with the fact that $\mathbf{P}$ is non-intersecting, implies that the sign of $\sigma_{\mathbf{P}}$ is independent of $\mathbf{P}$.

Consider the lattice $\mathcal{L}$ defined just before the beginning of this proof. Weight by $1/2$ its edges corresponding to dimer positions weighted by $1/2$ in the region $W_N \begin{pmatrix} R_1 & \ldots & R_m & R_1' & \ldots & R_n' \\ v_1 & & v_m & v_1' & & v_n' \end{pmatrix}$. Weight all other edges of $\mathcal{L}$ by 1. Then the weight

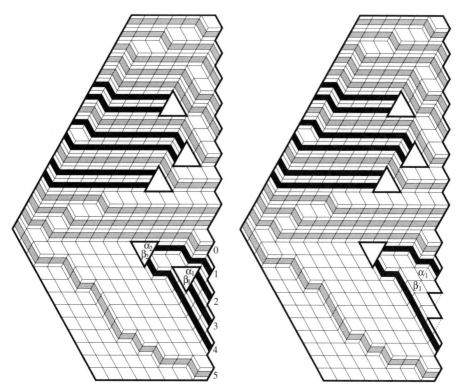

FIGURE 5.1. Lattice path encoding of
a tiling of $W_2 \begin{pmatrix} 5 & 2 & 4 & 2 & 3 \\ 0 & 1 & 1 & 2 & 4 \end{pmatrix}$.

FIGURE 5.2. The effect of Laplace expansion over the rows indexed by α and β.

of the dimer covering encoded by $\mathbf{P}$ is just $\mathrm{wt}(\mathbf{P})$, and we obtain by Theorem 5.2 and the constancy of the sign of $\sigma_{\mathbf{P}}$ that

$$(5.2) \qquad \mathrm{M}\left(W_N \begin{pmatrix} R_1 & \dots & R_m & R'_1 & \dots & R'_n \\ v_1 & & v_m & v'_1 & & v'_n \end{pmatrix}\right) = |\det A|,$$

where A is the $(2N+2m+4n+1) \times (2N+2m+4n+1)$ matrix recording the weighted counts of the lattice paths on $\mathcal{L}$ with given starting and ending points (note that the right hand side of (5.2) is independent of the ordering of these starting and ending points).

We deduce (5.1) by applying a sequence of Laplace expansions to the determinant in (5.2). Recall that for any $m \times m$ matrix M and any s-subset S of $[m] := \{1, \dots, m\}$, Laplace expansion along the rows with indices in S states that

$$(5.3) \qquad \det M = \sum_K (-1)^{\epsilon(K)} \det M_S^K \det M_{[m] \backslash S}^{[m] \backslash K},$$

where K ranges over all s-subsets of $[m]$, $\epsilon(K) := \sum_{k \in K} (k - 1)$ and M_I^J is the submatrix of M with row-index set I and column-index set J.

The rows and columns of the matrix A in (5.2) are indexed by the starting and ending points of the $(2N + 2m + 4n + 1)$-tuples of non-intersecting lattice paths encoding the dimer coverings of $W_N \begin{pmatrix} R_1 & \dots & R_m & R'_1 & \dots & R'_n \\ v_1 & & v_m & v'_1 & & v'_n \end{pmatrix}$. The starting points are the $2N + 4n + 1$ unit segments along the northwestern boundary of $W_N \begin{pmatrix} R_1 & \cdots & R_m \\ v_1 & & v_m \end{pmatrix} ; \begin{matrix} R'_1 & \cdots & R'_n \\ v'_1 & & v'_n \end{matrix}$, together with the segments α_i and β_i of D_i that are parallel to d, $i = 1, \ldots, m$ (see Figure 5.1). The ending points are the $2N+2m+2n+1$ segments parallel to d on the eastern boundary of $W_N \begin{pmatrix} R_1 & \cdots & R_m \\ v_1 & & v_m \end{pmatrix} ; \begin{matrix} R'_1 & \cdots & R'_n \\ v'_1 & & v'_n \end{matrix}$, together with two more such segments on each U_j, $j = 1, \ldots, n$.

Label the bottommost $N + 2m$ ending segments on the eastern boundary, from top to bottom, by $0, 1, \ldots, N + 2m - 1$ (this labeling is illustrated in Figure 5.1).

Apply Laplace expansion to the matrix A of (5.2) along the two rows indexed by α_1 and β_1 (see Figure 5.1). The first determinant in the summand in (5.3) is then just a two by two determinant. Its entries are weighted counts of lattice paths on $\mathcal{L}$ that start at α_1 or β_1 and end at some labeled segment on the eastern boundary. There are only $R_1 + 1$ labeled segments that can be reached this way: those with labels ranging from v_1 to $v_1 + R_1$. We can restrict summation in (5.3) to the two-element subsets K of this set of segments, since all other terms have at least one zero column in the two by two determinant. Therefore we obtain from (5.2) that

$$\mathrm{M}\left(W_N \begin{pmatrix} R_1 & \cdots & R_m \\ v_1 & & v_m \end{pmatrix} ; \begin{matrix} R'_1 & \cdots & R'_n \\ v'_1 & & v'_n \end{matrix} \right) =$$

(5.4)

$$\left| \sum_{0 \leq a_1 < b_1 \leq R_1} (-1)^{a_1+b_1} \det A_{\{\alpha_1,\beta_1\}}^{\{v_1+a_1,v_1+b_1\}} \det A_{[2N+2m+4n+1]\setminus\{\alpha_1,\beta_1\}}^{[2N+2m+4n+1]\setminus\{v_1+a_1,v_1+b_1\}} \right|$$

(here $[2N + 2m + 4n + 1]$ denotes the set of starting, respectively ending points of the families of non-intersecting lattice paths encoding tilings of the region (5.2)). Choosing the origin of the lattice $\mathcal{L}$ to be at α_1, and its positive axis directions to point east and southeast, one sees that β_1 has coordinates $(-1, 1)$, and the segment labeled $v_1 + j$ on the eastern boundary has coordinates $(R_1 - 1 - j, 2j)$, $j = 0, \ldots, R_1$. Since the lattice paths counted by the entries of $A_{\{\alpha_1,\beta_1\}}^{\{v_1+a_1,v_1+b_1\}}$ have all steps weighted by 1, the determinant of this matrix is

$$\det A_{\{\alpha_1,\beta_1\}}^{\{v_1+a_1,v_1+b_1\}} = \det \begin{bmatrix} \binom{R_1-1+a_1}{2a_1} & \binom{R_1-1+b_1}{2b_1} \\ \binom{R_1-1+a_1}{2a_1-1} & \binom{R_1-1+b_1}{2b_1-1} \end{bmatrix}$$

(5.5)
$$= 2R_1 \frac{(b_1 - a_1)(R_1 + a_1 - 1)! \, (R_1 + b_1 - 1)!}{(2a_1)! \, (R_1 - a_1)! \, (2b_1)! \, (R_1 - b_1)!}.$$

On the other hand, the second determinant in the summand in (5.4) can be interpreted as being the weighted count of dimer coverings of the region

(5.6) $$W_N \begin{pmatrix} R_2 & \cdots & R_m \\ v_2 & & v_m \end{pmatrix} ; \begin{matrix} R'_1 & \cdots & R'_n \\ v'_1 & & v'_n \end{matrix} \begin{bmatrix} v_1 + a_1 \\ v_1 + b_1 \end{bmatrix}$$

obtained from $W_N \begin{pmatrix} R_1 & \cdots & R_m \\ v_1 & & v_m \end{pmatrix} ; \begin{matrix} R'_1 & \cdots & R'_n \\ v'_1 & & v'_n \end{matrix}$ by placing back quadromer D_1 and removing the two monomers that contain the segments labeled $v_1 + a_1$ and $v_1 + b_1$ on the eastern boundary (see Figure 5.2 for an illustration). Indeed, the Lindström-Gessel-Viennot matrix of this region is precisely $A_{[2N+2m+4n+1]\setminus\{\alpha_1,\beta_1\}}^{[2N+2m+4n+1]\setminus\{v_1+a_1,v_1+b_1\}}$, and by the argument that proved (5.2) we obtain that

$$
\mathrm{M}\left(W_N \begin{pmatrix} R_2 & \cdots & R_m & R'_1 & \cdots & R'_n \\ v_2 & & v_m & v'_1 & & v'_n \end{pmatrix} \begin{bmatrix} v_1 + a_1 \\ v_1 + b_1 \end{bmatrix}\right)
$$

is equal to $\det A^{[2N+2m+4n+1]\setminus\{v_1+a_1, v_1+b_1\}}_{[2N+2m+4n+1]\setminus\{\alpha_1, \beta_1\}}$, up to a sign that is independent of a_1 and b_1 (to check the latter statement, in the labeling of the starting and ending points of $\mathbf{P}$ that defines $\sigma_{\mathbf{P}}$, label the bottommost $N + 2m - 2$ northwest facing unit segments on the boundary of (5.6), say from bottom to top, by $1, \ldots, N + 2m - 2$, irrespective of the values of a_1 and b_1; the argument in the second paragraph of the current proof shows that the sign of $\sigma_{\mathbf{P}}$ is independent of not just $\mathbf{P}$, but also of a_1 and b_1). Therefore, using (5.5) we can rewrite (5.4) as

$$
\mathrm{M}\left(W_N \begin{pmatrix} R_1 & \cdots & R_m & R'_1 & \cdots & R'_n \\ v_1 & & v_m & v'_1 & & v'_n \end{pmatrix}\right) =
$$

$$
2R_1 \left| \sum_{0 \le a_1 < b_1 \le R_1} (-1)^{a_1 + b_1} \frac{(b_1 - a_1)(R_1 + a_1 - 1)!\,(R_1 + b_1 - 1)!}{(2a_1)!\,(R_1 - a_1)!\,(2b_1)!\,(R_1 - b_1)!} \right.
$$

(5.7)

$$
\left. \times \mathrm{M}\left(W_N \begin{pmatrix} R_2 & \cdots & R_m & R'_1 & \cdots & R'_n \\ v_2 & & v_m & v'_1 & & v'_n \end{pmatrix} \begin{bmatrix} v_1 + a_1 \\ v_1 + b_1 \end{bmatrix}\right) \right|.
$$

This way the matching generating function of $\mathrm{M}\left(W_N \begin{pmatrix} R_1 & \cdots & R_m & R'_1 & \cdots & R'_n \\ v_1 & & v_m & v'_1 & & v'_n \end{pmatrix}\right)$ is expressed as a double sum involving the matching generating functions of regions having one less down-pointing triangular hole. By the same reasoning the second down-pointing quadromer D_2 can be removed from the regions (5.6) and the last factor in the summand of (5.7) is expressed as a double sum involving regions with *two* less down-pointing holes. One obtains

$$
\mathrm{M}\left(W_N \begin{pmatrix} R_1 & \cdots & R_m & R'_1 & \cdots & R'_n \\ v_1 & & v_m & v'_1 & & v'_n \end{pmatrix}\right) = 4R_1 R_2
$$

$$
\times \left| \sum_{0 \le a_1 < b_1 \le R_1} \sum_{0 \le a_2 < b_2 \le R_2} (-1)^{a_1 + b_1 + a_2 + b_2} \operatorname{sgn}(v_1 + a_1, v_1 + b_1, v_2 + a_2, v_2 + b_2) \right.
$$

$$
\times \frac{(b_1 - a_1)(R_1 + a_1 - 1)!\,(R_1 + b_1 - 1)!}{(2a_1)!\,(R_1 - a_1)!\,(2b_1)!\,(R_1 - b_1)!}
$$

$$
\times \frac{(b_2 - a_2)(R_2 + a_2 - 1)!\,(R_2 + b_2 - 1)!}{(2a_2)!\,(R_2 - a_2)!\,(2b_2)!\,(R_2 - b_2)!}
$$

(5.8)

$$
\left. \times \mathrm{M}\left(W_N \begin{pmatrix} R_3 & \cdots & R_m & R'_1 & \cdots & R'_n \\ v_3 & & v_m & v'_1 & & v'_n \end{pmatrix} \begin{bmatrix} v_1 + a_1 & v_2 + a_2 \\ v_1 + b_1 & v_2 + b_2 \end{bmatrix}\right) \right|,
$$

where the summation extends only over those indices for which $(v_1 + a_1, v_1 + b_1, v_2 + a_2, v_2 + b_2)$ consists of distinct components, sgn denotes its sign when regarded as a permutation, and the region on the right hand side of (5.8) is obtained from (5.6) by placing back the quadromer D_2 and removing the two monomers containing the segments labeled $v_2 + a_2$ and $v_2 + b_2$ on its eastern boundary.

Indeed, perform Laplace expansion for the region (5.6) over the rows indexed by α_2 and β_2. Because of the geometry of the triangular lattice $\mathcal{T}$, the paths starting at these segments can end on the eastern boundary only at segments with labels in the range $[v_2, v_2 + R_2]$; denote these labels by $v_2 + a_2$ and $v_2 + b_2$. Moreover, because the segments labeled $v_1 + a_1$ and $v_1 + b_1$ are not present in the region (5.6), the labels $v_2 + a_2$ and $v_2 + b_2$ must in fact also be different from both $v_1 + a_1$ and $v_1 + b_1$. This explains why the summation range in (5.8) needs to be restricted to $v_1 + a_1, v_1 + b_1, v_2 + a_2, v_2 + b_2$ being distinct.

The only other needed explanation for justifying (5.8) is the factor $\mathrm{sgn}(v_1 + a_1, v_1 + b_1, v_2 + a_2, v_2 + b_2)$. The reason it appears is the following. In the summand of the Laplace expansion (5.3) one has the factor $(-1)^{\epsilon(K)}$. In our situation this becomes $(-1)^{k_1 + k_2}$, where the "column indices" k_i indicate that the two path-ending segments on the eastern boundary are the k_1th and k_2th from, say, the top, *in the current region*. This column index coincides—up to a translation, which, pertaining to both k_1 and k_2, does not affect the sign $(-1)^{\epsilon(K)}$—in the case of $\mathrm{M}\left(W_N\begin{pmatrix} R_1 & \ldots & R_m & R_1' & \ldots & R_n' \\ v_1 & & v_m & v_1' & & v_n' \end{pmatrix}\right)$, with the label of the path-ending segment. However, when we do Laplace expansion for the region (5.6), this column index is affected—unless $v_2 + a_2 < v_2 + b_2 < v_1 + a_1 < v_1 + b_1$— by the previous removal of the segments with labels $v_1 + a_1$ and $v_1 + b_1$.

More precisely, say $v_2 + a_2 < v_1 + a_1 < v_2 + b_2 < v_1 + b_1$. Then while $v_2 + a_2$ correctly gives the column index of the segment where the path starting at α_2 ends, $v_2 + b_2$ is, due to the absence of the segment labeled $v_1 + a_1$, *one unit more* than the column index of the ending segment of the path starting at β_2. It is easy to see that in general the effect of this "interference" is precisely the multiplication of the summand by $\mathrm{sgn}(v_2 + a_2, v_2 + b_2, v_1 + a_1, v_1 + b_1)$. Since clearly $\mathrm{sgn}(v_2 + a_2, v_2 + b_2, v_1 + a_1, v_1 + b_1) = \mathrm{sgn}(v_1 + a_1, v_1 + b_1, v_2 + a_2, v_2 + b_2)$, (5.8) is completely justified.

Applying the same reasoning $m - 2$ more times, we obtain

$$
\mathrm{M}\left(W_N\begin{pmatrix} R_1 & \ldots & R_m & R_1' & \ldots & R_n' \\ v_1 & & v_m & v_1' & & v_n' \end{pmatrix}\right) = 2^m \prod_{i=1}^{m} R_i
$$

$$
\times \left| \sum_{0 \leq a_1 < b_1 \leq R_1} \cdots \sum_{0 \leq a_m < b_m \leq R_m} (-1)^{\sum_{i=1}^{m}(a_i + b_i)} \right.
$$

$$
\times \, \mathrm{sgn}(v_1 + a_1, v_1 + b_1, \ldots, v_m + a_m, v_m + b_m)
$$

$$
\times \prod_{i=1}^{m} \frac{(b_i - a_i)(R_i + a_i - 1)! \, (R_i + b_i - 1)!}{(2a_i)! \, (R_i - a_i)! \, (2b_i)! \, (R_i - b_i)!}
$$

(5.9)

$$
\times \left. \mathrm{M}\left(W_N\begin{pmatrix} \emptyset & R_1' & \ldots & R_n' \\ \emptyset & v_1' & & v_n' \end{pmatrix}\begin{bmatrix} v_1 + a_1 & \ldots & v_m + a_m \\ v_1 + b_1 & & v_m + b_m \end{bmatrix}\right) \right|,
$$

where the region on the right hand side of (5.9) is obtained from the region on the left by placing back all m down-pointing quadromers, and removing instead the $2m$ monomers containing the segments with labels $v_i + a_i$ and $v_i + b_i$, $i = 1, \ldots, m$, on its eastern boundary.

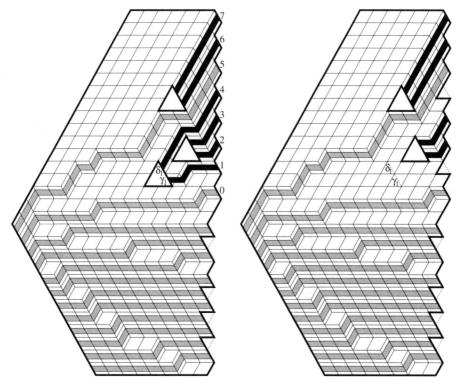

FIGURE 5.3. Lattice path encoding of a tiling of $W_2 \begin{pmatrix} \emptyset & 4 & 2 & 3 \\ \emptyset & 1 & 2 & 4 \end{pmatrix} \begin{bmatrix} 0 & 2 \\ 4 & 3 \end{bmatrix}$.

FIGURE 5.4. The effect of Laplace expansion over the rows indexed by γ and δ.

We complete expressing the left hand side of (5.1) in terms of regions with no holes by repeating our hole-removing procedure for the regions

$$(5.10) \qquad W_N \begin{pmatrix} \emptyset & R'_1 & \dots & R'_n \\ \emptyset & v'_1 & & v'_n \end{pmatrix} \begin{bmatrix} v_1 + a_1 & \dots & v_m + a_m \\ v_1 + b_1 & & v_m + b_m \end{bmatrix}$$

on the right hand side of (5.9).

To this end, we encode the tilings of the regions (5.10) by paths of dimers parallel to the direction d now chosen to be the southeast-northwest direction (an example is illustrated in Figure 5.3). As in the previous encoding, weight by $1/2$ those segments of the resulting lattice $\mathcal{L}$ that correspond to dimer positions weighted $1/2$ in the region (5.10), and weight all remaining segments of $\mathcal{L}$ by 1.

Each tiling of the region (5.10) gets encoded this way by a $(2N + 4m + 2n)$-tuple $\mathbf{P}$ of non-intersecting lattice paths on $\mathcal{L}$, starting at the unit segments on its southwestern boundary or at the unit segments γ_i and δ_i of U_i $(i = 1, \dots, n)$ that are parallel to d, and ending at the unit segments parallel to d on its eastern boundary (see Figure 5.3).

Label the topmost $N + 2n$ ending segments on the eastern boundary, from bottom to top, by $0, 1, \dots, N + 2n - 1$ (this labeling is illustrated in Figure 5.3).

As in the argument that proved (5.2), the sign of the permutation $\sigma_{\mathbf{P}}$ is independent of $\mathbf{P}$. Therefore, we obtain by Theorem 5.2 that

$$(5.11) \qquad \mathrm{M}\left(W_N\begin{pmatrix}\emptyset, & R_1' & \cdots & R_n' \\ \emptyset; & v_1' & & v_n'\end{pmatrix}\begin{bmatrix}v_1+a_1 & \cdots & v_m+a_m \\ v_1+b_1 & & v_m+b_m\end{bmatrix}\right) = \epsilon_1 \det B,$$

where B is the $(2N+4m+2n)\times(2N+4m+2n)$ matrix recording the weighted counts of the lattice paths with specified starting and ending points, and the sign ϵ_1 in front of the determinant is the same for all choices of a_i and b_i, $i=1,\ldots,m$.

Apply Laplace expansion in $\det B$ along the two rows indexed by γ_1 and δ_1. The first determinant in the summand of (5.3) is again two by two, and records weighted counts of lattice paths starting at γ_1 or δ_1 and ending at some segment on the eastern boundary (see Figure 5.3). There are $R_1'+1$ segments on the eastern boundary that can be reached this way; in the labeling of the paragraph before (5.11), they are the ones with labels $v_1', v_1'+1, \ldots, v_1'+R_1'$. As with our previous Laplace expansions, we can restrict the summation range in (5.3) to obtain

$$\mathrm{M}\left(W_N\begin{pmatrix}\emptyset, & R_1' & \cdots & R_n' \\ \emptyset; & v_1' & & v_n'\end{pmatrix}\begin{bmatrix}v_1+a_1 & \cdots & v_m+a_m \\ v_1+b_1 & & v_m+b_m\end{bmatrix}\right) =$$

(5.12)

$$\epsilon_1 \sum_{0\le c_1<d_1\le R_1'} (-1)^{c_1+d_1} \det B_{\{\gamma_1,\delta_1\}}^{\{v_1'+c_1,v_1'+d_1\}} \det B_{[2N+4m+2n]\backslash\{\gamma_1,\delta_1\}}^{[2N+4m+2n]\backslash\{v_1'+c_1,v_1'+d_1\}}.$$

Centering $\mathcal{L}$ at γ_1 and choosing the positive directions in the lattice $\mathcal{L}$ to point east and northeast, δ_1 has coordinates $(-1,1)$, and the segment labeled $v_1'+j$ on the eastern boundary has coordinates $(R_1'-1-j, 2j+1)$, $j=0,1,\ldots,R_1'$. The weighted counts involved in the entries of $B_{\{\gamma_1,\delta_1\}}^{\{v_1'+c_1,v_1'+d_1\}}$ (which involve this time some steps of weight $1/2$) are easily calculated and one obtains

$$\det B_{\{\gamma_1,\delta_1\}}^{\{v_1'+c_1,v_1'+d_1\}} = \det \begin{bmatrix} \frac{1}{2}\binom{R_1'-1+c_1}{2c_1}+\binom{R_1'-1+c_1}{2c_1+1} & \frac{1}{2}\binom{R_1'-1+d_1}{2d_1}+\binom{R_1'-1+d_1}{2d_1+1} \\ \frac{1}{2}\binom{R_1'-1+c_1}{2c_1-1}+\binom{R_1'-1+c_1}{2c_1} & \frac{1}{2}\binom{R_1'-1+d_1}{2d_1-1}+\binom{R_1'-1+d_1}{2d_1} \end{bmatrix}$$

$$= 2R_1'(R_1'-1/2)(R_1'+1/2)\frac{(d_1-c_1)(R_1'+c_1-1)!\,(R_1'+d_1-1)!}{(2c_1+1)!\,(R_1'-c_1)!\,(2d_1+1)!\,(R_1'-d_1)!}.$$

(5.13)

On the other hand, by applying Theorem 5.2 one more time one sees that

$$\det B_{[2N+4m+2n]\backslash\{\gamma_1,\delta_1\}}^{[2N+4m+2n]\backslash\{v_1'+c_1,v_1'+d_1\}} =$$

(5.14)

$$\epsilon_1' \,\mathrm{M}\left(W_N\begin{pmatrix}\emptyset, & R_2' & \cdots & R_n' \\ \emptyset; & v_2' & & v_n'\end{pmatrix}\begin{bmatrix}v_1+a_1 & \cdots & v_m+a_m \\ v_1+b_1 & & v_m+b_m\end{bmatrix}\begin{bmatrix}v_1'+c_1 \\ v_1'+d_1\end{bmatrix}\right),$$

where

$$(5.15) \qquad W_N\begin{pmatrix}\emptyset, & R_2' & \cdots & R_n' \\ \emptyset; & v_2' & & v_n'\end{pmatrix}\begin{bmatrix}v_1+a_1 & \cdots & v_m+a_m \\ v_1+b_1 & & v_m+b_m\end{bmatrix}\begin{bmatrix}v_1'+c_1 \\ v_1'+d_1\end{bmatrix}$$

is the region obtained from the region (5.10) by placing back quadromer U_1 and removing the two monomers containing the segments labeled $v_1'+c_1$ and $v_1'+d_1$

on the eastern boundary, and the sign ϵ_1' is independent of c_1 and d_1. Therefore, by (5.12)–(5.14) we obtain that

$$
\mathrm{M}\left(W_N \begin{pmatrix} \emptyset, & R_1' & \cdots & R_n' \\ \emptyset \,; & v_1' & & v_n' \end{pmatrix} \begin{bmatrix} v_1 + a_1 & \cdots & v_m + a_m \\ v_1 + b_1 & & v_m + b_m \end{bmatrix}\right) = 2R_1'(R_1' - 1/2)(R_1' + 1/2)
$$

$$
\times \epsilon_1 \epsilon_1' \sum_{0 \le c_1 < d_1 \le R_1'} (-1)^{c_1 + d_1} \frac{(d_1 - c_1)(R_1' + c_1 - 1)!\,(R_1' + d_1 - 1)!}{(2c_1 + 1)!\,(R_1' - c_1)!\,(2d_1 + 1)!\,(R_1' - d_1)!}
$$

(5.16)

$$
\times \mathrm{M}\left(W_N \begin{pmatrix} \emptyset, & R_2' & \cdots & R_n' \\ \emptyset \,; & v_2' & & v_n' \end{pmatrix} \begin{bmatrix} v_1 + a_1 & \cdots & v_m + a_m , & v_1' + c_1 \\ v_1 + b_1 & & v_m + b_m , & v_1' + d_1 \end{bmatrix}\right).
$$

By further Laplace expansions applied on the pairs of rows indexed by γ_i and δ_i, $i = 2, \ldots, n$, one can successively remove all remaining holes $U_2, \ldots, U_n$. The argument that proved (5.9) yields now

$$
\mathrm{M}\left(W_N \begin{pmatrix} \emptyset, & R_1' & \cdots & R_n' \\ \emptyset \,; & v_1' & & v_n' \end{pmatrix} \begin{bmatrix} v_1 + a_1 & \cdots & v_m + a_m \\ v_1 + b_1 & & v_m + b_m \end{bmatrix}\right) = \prod_{i=1}^{n} \epsilon_i \epsilon_i'
$$

$$
\times \prod_{i=1}^{n} 2R_i'(R_i' - 1/2)(R_i' + 1/2)
$$

$$
\times \sum_{0 \le c_1 < d_1 \le R_1'} \cdots \sum_{0 \le c_n < d_n \le R_n'} (-1)^{\sum_{i=1}^{n}(c_i + d_i)}
$$

$$
\times \mathrm{sgn}(v_1' + c_1, v_1' + d_1, \ldots, v_n' + c_n, v_n' + d_n)
$$

$$
\times \prod_{i=1}^{n} \frac{(d_i - c_i)(R_i' + c_i - 1)!\,(R_i' + d_i - 1)!}{(2c_i + 1)!\,(R_i' - c_i)!\,(2d_i + 1)!\,(R_i' - d_i)!}
$$

(5.17)

$$
\times \mathrm{M}\left(W_N \begin{pmatrix} \emptyset, & \emptyset \\ \emptyset \,; & \emptyset \end{pmatrix} \begin{bmatrix} v_1 + a_1 & \cdots & v_m + a_m , & v_1' + c_1 & \cdots & v_n' + c_n \\ v_1 + b_1 & & v_m + b_m , & v_1' + d_1 & & v_n' + d_n \end{bmatrix}\right),
$$

where the multiple sum extends only to the summation indices for which $v_1' + c_1, v_1' + d_1, \ldots, v_n' + c_n, v_n' + d_n$ are distinct, the signs ϵ_i and ϵ_i', $i = 1, \ldots, n$, are independent of $a_1, b_1, \ldots, a_m, b_m$, and

$$
(5.18) \qquad W_N \begin{pmatrix} \emptyset, & \emptyset \\ \emptyset \,; & \emptyset \end{pmatrix} \begin{bmatrix} v_1 + a_1 & \cdots & v_m + a_m , & v_1' + c_1 & \cdots & v_n' + c_n \\ v_1 + b_1 & & v_m + b_m , & v_1' + d_1 & & v_n' + d_n \end{bmatrix}
$$

is the region obtained from the region (5.10) by placing back all its removed quadromers U_i, $i = 1, \ldots, n$, and removing the $2n$ monomers containing the segments labeled γ_i and δ_i, $i = 1, \ldots, n$, on its eastern boundary.

However, the region (5.18) differs from the region $W_N[\{v_1 + a_1, v_1 + b_1, \ldots, v_m + a_m, v_m + b_m\}_<; \{v_1' + c_1, v_1' + d_1, \ldots, v_n' + c_n, v_n' + d_n\}_<]$ (where $A_<$ denotes the list of increasingly sorted elements of the set A of distinct nonnegative integers) defined in Section 4 only in that the former contains $2m + 2n$ more unit rhombi near the eastern boundary (see Figure 5.5; the additional rhombi are shaded). Moreover, all these additional dimer positions have weight 1 and are forced to be part of any dimer covering of the former region. Therefore the two regions have the same matching generating function, and replacing (5.17) in the right hand side of (5.9)

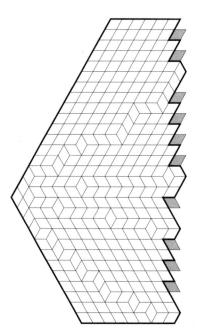

FIGURE 5.5. The regions $W_N \begin{pmatrix} \emptyset, & \emptyset \\ \emptyset ; & \emptyset \end{pmatrix} \begin{bmatrix} 0 & 2, 0 & 1 & 5 \\ 4 & 3 ; 3 & 2 & 6 \end{bmatrix}$ and $W_N[0,2,3,4;0,1,2,3,5,6]$ have the same matching generating function.

one obtains

$$
\mathrm{M}\left(W_N\begin{pmatrix} R_1 & \dots & R_m ; & R_1' & \dots & R_n' \\ v_1 & & v_m ; & v_1' & & v_n' \end{pmatrix}\right) = 2^{m+n} \prod_{i=1}^m R_i \prod_{i=1}^n R_i'(R_i'-1/2)(R_i'+1/2)
$$

$$
\times \Bigg| \sum_{0 \le a_1 < b_1 \le R_1} \cdots \sum_{0 \le a_m < b_m \le R_m} \sum_{0 \le c_1 < d_1 \le R_1'} \cdots \sum_{0 \le c_n < d_n \le R_n'} (-1)^{\sum_{i=1}^m (a_i+b_i)}
$$

$$
\times (-1)^{\sum_{i=1}^n (c_i+d_i)}
$$

$$
\times \prod_{i=1}^m \frac{(b_i-a_i)(R_i+a_i-1)!\,(R_i+b_i-1)!}{(2a_i)!\,(R_i-a_i)!\,(2b_i)!\,(R_i-b_i)!}
$$

$$
\times \prod_{i=1}^n \frac{(d_i-c_i)(R_i'+c_i-1)!\,(R_i'+d_i-1)!}{(2c_i+1)!\,(R_i'-c_i)!\,(2d_i+1)!\,(R_i'-d_i)!}
$$

$$
\times \mathrm{sgn}(v_1'+c_1, v_1'+d_1, \dots, v_n'+c_n, v_n'+d_n)
$$

$$
\times \mathrm{sgn}(v_1+a_1, v_1+b_1, \dots, v_m+a_m, v_m+b_m)
$$

$$
\times \mathrm{M}(W_N[\{v_1+a_1, \dots, v_m+b_m\}_<; \{v_1'+c_1, \dots, v_n'+d_n\}_<]) \Bigg|,
$$

(5.19)

the summation extending only over those summation indices for which both lists of arguments in the region on the right hand side have distinct elements.

After removing the forced dimers, the normalizing region at the denominator of (2.8),

$$(5.20) \qquad W_N \begin{pmatrix} 1 & 3 & \cdots & 2m-1 \\ 0 & 0 & & 0 \end{pmatrix} ; \begin{matrix} 1 & 3 & \cdots & 2n-1 \\ 0 & 0 & & 0 \end{matrix} \end{pmatrix},$$

is seen to be precisely $W_N[0, 1, \ldots, 2m-1; 0, 1, \ldots, 2n-1]$.

Moreover, it follows from (4.2) that

$$\lim_{N \to \infty} \frac{\mathrm{M}\left(W_N[\{v_1 + a_1, \ldots, v_m + b_m\}_<; \{v_1' + c_1, \ldots, v_n' + d_n\}_<]\right)}{\mathrm{M}\left(W_N[0, \ldots, 2m-1; 0, \ldots, 2n-1]\right)} = \chi_{2m,2n}$$

$$\times \, \mathrm{sgn}(v_1' + c_1, v_1' + d_1, \ldots, v_n' + c_n, v_n' + d_n)$$

$$\times \, \mathrm{sgn}(v_1 + a_1, v_1 + b_1, \ldots, v_m + a_m, v_m + b_m)$$

$$\times \prod_{i=1}^{m} \frac{(3/2)_{v_i+a_i}\, (3/2)_{v_i+b_i}}{(2)_{v_i+a_i}\, (2)_{v_i+b_i}} \prod_{i=1}^{n} \frac{(3/2)_{v_i'+c_i}\, (3/2)_{v_i'+d_i}}{(1)_{v_i'+c_i}\, (1)_{v_i'+d_i}}$$

$$\times \prod_{1 \le i < j \le m} (v_j - v_i + a_j - a_i)(v_j - v_i + a_j - b_i)$$

$$\times \, (v_j - v_i + b_j - a_i)(v_j - v_i + b_j - b_i)$$

$$\times \prod_{1 \le i < j \le n} (v_j' - v_i' + c_j - c_i)(v_j' - v_i' + c_j - d_i)$$

$$\times \, (v_j' - v_i' + d_j - c_i)(v_j' - v_i' + d_j - d_i)$$

$$(5.21)$$

$$\times \frac{\prod_{i=1}^{m}(a_i - b_i) \prod_{i=1}^{n}(c_i - d_i)}{\prod_{i=1}^{m} \prod_{j=1}^{n}(u_{ij} + a_i + c_j)(u_{ij} + a_i + d_j)(u_{ij} + b_i + c_j)(u_{ij} + b_i + d_j)},$$

with χ given by (4.3) and $u_{ij} = v_i + v_j' + 2$, $i = 1, \ldots, m$, $j = 1, \ldots, n$. Indeed, (5.21) is a direct consequence of (4.2) when $v_1 + a_1 < v_1 + b_1 < \cdots < v_m + a_m < v_m + b_m$ and $v_1' + c_1 < v_1' + d_1 < \cdots < v_n' + c_n < v_n' + d_n$. Since each "elementary move" changing the relative order of just two consecutive elements in either of these two lists (except when these two elements are $v_i + a_i$ and $v_i + b_i$, or $v_i' + c_i$ and $v_i' + d_i$, for some i, case in which their relative order is fixed by construction) has the effect of multiplying *both* the sign part and the product part of the right hand side of (5.21) by -1, it follows that (5.21) is valid in general.

Therefore, divide (5.19) by the matching generation function of the region (5.20) and let $N \to \infty$. By (2.8), the left hand side becomes the left hand side of (5.1), while by (5.21) the summand on the right hand side becomes precisely the summand on the right hand side of (5.1). (In particular, it follows that the limit (2.6) defining

ω_b exists.) We obtain

$$\omega_b \begin{pmatrix} R_1 \ \ldots \ R_m \ ; \ R_1' \ \ldots \ R_n' \\ v_1 \ \ldots \ v_m \ ; \ v_1' \ \ldots \ v_n' \end{pmatrix} = 2^{m+n} \chi_{m,n} \prod_{i=1}^{m} R_i \prod_{i=1}^{n} R_i'(R_i' - 1/2)(R_i' + 1/2)$$

$$\times \Bigg| \sum_{0 \le a_1 < b_1 \le R_1} \cdots \sum_{0 \le a_m < b_m \le R_m} \sum_{0 \le c_1 < d_1 \le R_1'} \cdots \sum_{0 \le c_n < d_n \le R_n'} (-1)^{\sum_{i=1}^m (a_i + b_i) + \sum_{i=1}^n (c_i + d_i)}$$

$$\times \prod_{i=1}^{m} \frac{(R_i + a_i - 1)! \, (R_i + b_i - 1)!}{(2a_i)! \, (R_i - a_i)! \, (2b_i)! \, (R_i - b_i)!} \frac{(3/2)_{v_i + a_i} \, (3/2)_{v_i + b_i}}{(2)_{v_i + a_i} \, (2)_{v_i + b_i}}$$

$$\times \prod_{i=1}^{n} \frac{(R_i + c_i - 1)! \, (R_i + d_i - 1)!}{(2c_i + 1)! \, (R_i - c_i)! \, (2d_i + 1)! \, (R_i - d_i)!} \frac{(3/2)_{v_i' + c_i} \, (3/2)_{v_i' + d_i}}{(1)_{v_i' + c_i} \, (1)_{v_i' + d_i}}$$

$$\times \prod_{1 \le i < j \le m} (v_j - v_i + a_j - a_i)(v_j - v_i + a_j - b_i)(v_j - v_i + b_j - a_i)(v_j - v_i + b_j - b_i)$$

$$\times \prod_{1 \le i < j \le n} (v_j' - v_i' + c_j - c_i)(v_j' - v_i' + c_j - d_i)(v_j' - v_i' + d_j - c_i)(v_j' - v_i' + d_j - d_i)$$

$$\times \frac{\prod_{i=1}^{m} (a_i - b_i)^2 \prod_{i=1}^{n} (c_i - d_i)^2}{\prod_{i=1}^{m} \prod_{j=1}^{n} (u_{ij} + a_i + c_j)(u_{ij} + a_i + d_j)(u_{ij} + b_i + c_j)(u_{ij} + b_i + d_j)} \Bigg| ,$$

(5.22)

where the summation range is restricted to those summation variables for which $v_1 + a_1, v_1 + b_1, \ldots, v_m + a_m, v_m + b_m$, as well as $v_1' + c_1, v_1' + d_1, \ldots, v_n' + c_n, v_n' + d_n$, are distinct.

The fortunate situation is that, on the one hand, when $a_i = b_i$, for some $i = 1, \ldots, m$, or when $c_i = d_i$, for some $i = 1, \ldots, n$, the summand in (5.22) becomes zero. On the other, the summand in (5.22) is invariant under independently interchanging a_i with b_i, $i = 1, \ldots, m$, and c_i with d_i, $i = 1, \ldots, m$ (because the differences $b_i - a_i$ and $d_i - c_i$ end up, by the combination of (5.5), (5.13) and (5.21), appearing at the second power). Therefore the summation range may be extended in (5.22), at the expense of a multiplicative factor of $1/2^{m+n}$, to the summation range in (5.1). This leads precisely to the multiple sum given in the statement of the Lemma. $\qquad \square$

Separation of the $(2m + 2n)$-fold sum
for ω_b in terms of $4mn$-fold integrals

If it weren't for the denominator in the last fraction on the right hand side of (5.1), one could expand the product at the numerator of the summand in terms of monomials in the summation variables, and the $(2m + 2n)$-fold sum (5.1) would separate: one could express it in terms of simple sums.

We can get around this obstacle by expressing each factor at the denominator as an integral:

$$\frac{1}{u_{ij} + a_i + c_j} = \int_0^1 x_{ij}^{u_{ij}+a_i+c_j-1} dx_{ij}$$

$$\frac{1}{u_{ij} + a_i + d_j} = \int_0^1 y_{ij}^{u_{ij}+a_i+d_j-1} dy_{ij}$$

$$\frac{1}{u_{ij} + b_i + c_j} = \int_0^1 z_{ij}^{u_{ij}+b_i+c_j-1} dz_{ij}$$

(6.1)
$$\frac{1}{u_{ij} + b_i + d_j} = \int_0^1 w_{ij}^{u_{ij}+b_i+d_j-1} dw_{ij},$$

for $i = 1, \ldots, m$, $j = 1, \ldots, n$. Then the multiple sum inside the absolute value signs in (5.1) can be expressed as a sum of $4mn$-fold integrals of products of simple sums. More precisely, expand the product of the numerators on the last three lines of (5.1) as

$$\mathcal{E} := \prod_{i=1}^m (a_i - b_i)^2 \prod_{i=1}^n (c_i - d_i)^2$$

$$\times \prod_{1\leq i<j\leq m} ((v_j - v_i) + a_j - a_i)((v_j - v_i) + a_j - b_i)$$

$$\times ((v_j - v_i) + b_j - a_i)((v_j - v_i) + b_j - b_i)$$

$$\times \prod_{1\leq i<j\leq n} ((v_j' - v_i') + c_j - c_i)((v_j' - v_i') + c_j - d_i)$$

$$\times ((v_j' - v_i') + d_j - c_i)((v_j' - v_i') + d_j - d_i)$$

(6.2)
$$= \sum_{C\in\mathcal{C}} e(C) a_1^{\alpha_1(C)} b_1^{\beta_1(C)} \cdots a_m^{\alpha_m(C)} b_m^{\beta_m(C)} c_1^{\gamma_1(C)} d_1^{\delta_1(C)} \cdots c_n^{\gamma_n(C)} d_n^{\delta_n(C)},$$

where $\mathcal{C}$ is the collection of all $2^{2m+2n} 3^{4\binom{m}{2}+4\binom{n}{2}}$ signed monomials in the (v_j-v_i)'s, $(v_j' - v_i')$'s, a_i's, b_i's, c_j's and d_j's obtained by expanding the left hand side of (6.2), and for such a monomial $C \in \mathcal{C}$, $\alpha_i(C)$, $\beta_i(C)$, $\gamma_j(C)$ and $\delta_j(C)$ are the exponents of a_i, b_i, c_j and d_j, respectively, while $e(C)$ is the part of C besides $\prod_{i=1}^m a_i^{\alpha_i(C)} b_i^{\beta_i(C)} \prod_{j=1}^n c_j^{\gamma_j(C)} d_j^{\delta_j(C)}$ (so $e(C)$ is a signed monomial in the $(v_j - v_i)$'s and $(v_j' - v_i')$'s).

Define also

$$(6.3) \qquad T^{(n)}(R, v; x) := \frac{1}{R} \sum_{a=0}^{R} \frac{(-R)_a (R)_a (3/2)_{v+a}}{(1)_a (1/2)_a (2)_{v+a}} \left(\frac{x}{4}\right)^a a^n$$

$$(6.4) \qquad T'^{(n)}(R, v; x) := \frac{1}{R} \sum_{c=0}^{R} \frac{(-R)_c (R)_c (3/2)_{v+c}}{(1)_c (3/2)_c (1)_{v+c}} \left(\frac{x}{4}\right)^c c^n.$$

We have the following result.

PROPOSITION 6.1. *The boundary-influenced correlation ω_b can be written as*

$$\omega_b \begin{pmatrix} R_1 & \dots & R_m & R_1' & \dots & R_n' \\ v_1 & & v_m & v_1' & & v_n' \end{pmatrix} = \chi_{2m,2n} \prod_{i=1}^m R_i \prod_{i=1}^n R_i'(R_i' - 1/2)(R_i' + 1/2)$$

$$(6.5)$$

$$\times \left| \sum_{C \in \mathcal{C}} e(C) M_{\alpha_1(C), \beta_1(C), \dots, \alpha_m(C), \beta_m(C); \gamma_1(C), \delta_1(C), \dots, \gamma_n(C), \delta_n(C)} \right|,$$

where χ is given by (4.3), the collection $\mathcal{C}$ and $e(C)$, $\alpha_i(C)$, $\beta_i(C)$, $\gamma_j(C)$, $\delta_j(C)$ are as in (6.2), and the "moment" $M_{\alpha_1, \beta_1, \dots, \alpha_m, \beta_m; \gamma_1, \delta_1, \dots, \gamma_n, \delta_n}$ equals the $4mn$-fold integral

$$M_{\alpha_1, \beta_1, \dots, \alpha_m, \beta_m; \gamma_1, \delta_1, \dots, \gamma_n, \delta_n} = \int_0^1 \cdots \int_0^1 \prod_{i=1}^m \prod_{j=1}^n (x_{ij} y_{ij} z_{ij} w_{ij})^{v_i + v_j' + 1}$$

$$\times T^{(\alpha_1)}(R_1, v_1; \prod_{j=1}^n x_{1j} y_{1j}) \, T^{(\beta_1)}(R_1, v_1; \prod_{j=1}^n z_{1j} w_{1j}) \cdots$$

$$\cdots T^{(\alpha_m)}(R_m, v_m; \prod_{j=1}^n x_{mj} y_{mj}) \, T^{(\beta_m)}(R_m, v_m; \prod_{j=1}^n z_{mj} w_{mj})$$

$$\times T'^{(\gamma_1)}(R_1', v_1'; \prod_{i=1}^m x_{i1} z_{i1}) \, T'^{(\delta_1)}(R_1', v_1'; \prod_{i=1}^m y_{i1} w_{i1}) \cdots$$

$$(6.6)$$

$$\cdots T'^{(\gamma_n)}(R_n', v_n'; \prod_{i=1}^m x_{in} z_{in}) \, T'^{(\delta_n)}(R_n', v_n'; \prod_{i=1}^m y_{in} w_{in}) \, dx_{11} \cdots dw_{mn},$$

with $T^{(n)}(R, v; x)$ and $T'^{(n)}(R, v; x)$ defined by (6.3)–(6.4).

PROOF. We note first that, since

$$\frac{(R + a - 1)!}{(2a)! (R - a)!} = \frac{(-1)^a}{R} \frac{(-R)_a (R)_a}{4^a (1)_a (1/2)_a}$$

and

$$\frac{(R + c - 1)!}{(2c + 1)! (R - c)!} = \frac{(-1)^c}{R} \frac{(-R)_c (R)_c}{4^c (1)_c (3/2)_c},$$

the sums in (6.3)–(6.4) can also be written as

$$(6.7) \qquad T^{(n)}(R,v;x) = \sum_{a=0}^{R} (-1)^a \frac{(R+a-1)!}{(2a)!\,(R-a)!} \frac{(3/2)_{v+a}}{(2)_{v+a}} x^a a^n$$

$$(6.8) \qquad T'^{(n)}(R,v;x) = \sum_{c=0}^{R} (-1)^c \frac{(R+c-1)!}{(2c+1)!\,(R-c)!} \frac{(3/2)_{v+c}}{(1)_{v+c}} x^c c^n.$$

Denote by S the multiple sum in (5.1). Express the factors at the denominator of the summand of S as integrals, using (6.1). Expand the factors of the summand of S contained in the left hand side of (6.2) as a sum of monomials in the summation variables $a_1, b_1, \ldots, a_m, b_m$ and $c_1, d_1, \ldots, c_n, d_n$ of S, as indicated by (6.2). Bring all $4mn$ integration signs in front of S. We obtain

$$S = \int_0^1 \cdots \int_0^1 \left\{ \sum_{a_1,b_1=0}^{R_1} \cdots \sum_{a_m,b_m=0}^{R_m} \sum_{c_1,d_1=0}^{R'_1} \cdots \sum_{c_n,d_n=0}^{R'_n} (-1)^{\sum_{i=1}^m (a_i+b_i)+\sum_{i=1}^n (c_i+d_i)} \right.$$

$$\times \prod_{i=1}^m \frac{(R_i+a_i-1)!\,(R_i+b_i-1)!}{(2a_i)!\,(R_i-a_i)!\,(2b_i)!\,(R_i-b_i)!} \frac{(3/2)_{v_i+a_i}\,(3/2)_{v_i+b_i}}{(2)_{v_i+a_i}\,(2)_{v_i+b_i}}$$

$$\times \prod_{i=1}^n \frac{(R_i+c_i-1)!\,(R_i+d_i-1)!}{(2c_i+1)!\,(R_i-c_i)!\,(2d_i+1)!\,(R_i-d_i)!} \frac{(3/2)_{v'_i+c_i}\,(3/2)_{v'_i+d_i}}{(1)_{v'_i+c_i}\,(1)_{v'_i+d_i}}$$

$$\times \prod_{i=1}^m \prod_{j=1}^n x_{ij}^{v_i+v'_j+a_i+c_j+1}\, y_{ij}^{v_i+v'_j+a_i+d_j+1}\, z_{ij}^{v_i+v'_j+b_i+c_j+1}\, w_{ij}^{v_i+v'_j+b_i+d_j+1}$$

$$\times \sum_{C\in\mathcal{C}} e(C) a_1^{\alpha_1(C)} b_1^{\beta_1(C)} \cdots a_m^{\alpha_m(C)} b_m^{\beta_m(C)} c_1^{\gamma_1(C)} d_1^{\delta_1(C)} \cdots c_n^{\gamma_n(C)} d_n^{\delta_n(C)} \Bigg\}$$

$$dx_{11} \cdots dw_{mn}.$$

Reversing the order of the two multiple summations in the integrand above yields

$$S = \int_0^1 \cdots \int_0^1 \sum_{C\in\mathcal{C}} \left\{ \sum_{a_1,b_1=0}^{R_1} \cdots \sum_{a_m,b_m=0}^{R_m} \sum_{c_1,d_1=0}^{R'_1} \cdots \sum_{c_n,d_n=0}^{R'_n} (-1)^{\sum_{i=1}^m (a_i+b_i)+\sum_{i=1}^n (c_i+d_i)} \right.$$

$$\times \prod_{i=1}^m \frac{(R_i+a_i-1)!\,(R_i+b_i-1)!}{(2a_i)!\,(R_i-a_i)!\,(2b_i)!\,(R_i-b_i)!}$$

$$\times \prod_{i=1}^n \frac{(R_i+c_i-1)!\,(R_i+d_i-1)!}{(2c_i+1)!\,(R_i-c_i)!\,(2d_i+1)!\,(R_i-d_i)!}$$

$$\times \prod_{i=1}^m \frac{(3/2)_{v_i+a_i}\,(3/2)_{v_i+b_i}}{(2)_{v_i+a_i}\,(2)_{v_i+b_i}} \prod_{i=1}^n \frac{(3/2)_{v'_i+c_i}\,(3/2)_{v'_i+d_i}}{(1)_{v'_i+c_i}\,(1)_{v'_i+d_i}}$$

$$\times \prod_{i=1}^m \prod_{j=1}^n x_{ij}^{v_i+v'_j+a_i+c_j+1}\, y_{ij}^{v_i+v'_j+a_i+d_j+1}\, z_{ij}^{v_i+v'_j+b_i+c_j+1}\, w_{ij}^{v_i+v'_j+b_i+d_j+1}$$

$$\times a_1^{\alpha_1(C)} b_1^{\beta_1(C)} \cdots a_m^{\alpha_m(C)} b_m^{\beta_m(C)} c_1^{\gamma_1(C)} d_1^{\delta_1(C)} \cdots c_n^{\gamma_n(C)} d_n^{\delta_n(C)} \Bigg\} e(C)\, dx_{11} \cdots dw_{mn}.$$

Clearly, the inner multiple sum separates in terms of simple sums on the summation variables $a_1, b_1, \ldots, a_m, b_m$ and $c_1, d_1, \ldots, c_n, d_n$. Moreover, all the simple sums arising this way have one of the forms (6.7) or (6.8). Moving the $4mn$-fold integral sign inside the outer multiple sum above one obtains (6.6). $\qquad\square$

The sums (6.3) and (6.4) can conveniently be expressed in terms of hypergeometric functions[6]. Indeed, for integer R the upper summation limits in (6.3) and (6.4) can be replaced by ∞ without affecting the definitions, due to the factors $(-R)_a$ and $(-R)_c$ in the summands. Using that $(x)_{v+a} = (x)_v(x+v)_a$, (6.3) becomes

$$T^{(n)}(R,v;x) = \frac{1}{R}\frac{(3/2)_v}{(2)_v}\sum_{a=0}^{\infty}\frac{(-R)_a\,(R)_a\,(v+3/2)_a}{(1)_a\,(1/2)_a\,(v+2)_a}\left(\frac{x}{4}\right)^a a^n$$

$$= \frac{1}{R}\frac{(3/2)_v}{(2)_v}\sum_{a=0}^{\infty}\frac{(-R)_a\,(R)_a\,(v+3/2)_a}{(1)_a\,(1/2)_a\,(v+2)_a}\left(\frac{x}{4}\right)^a$$

$$\times\{f_n a(a-1)\cdots(a-n+1) + f_{n-1}a(a-1)\cdots(a-n+2) + \cdots + f_1 a + f_0\},$$

(6.9)

where the coefficients f_i are defined so that the last factor in the summand equals a^n (in particular, $f_n = 1$ and f_0 is the Kronecker symbol δ_{n0}). Since $a(a-1)\cdots(a-k+1)/(1)_a = 1/(1)_{a-k}$, for $a \geq k$, we have

$$\sum_{a=0}^{\infty}\frac{(-R)_a\,(R)_a\,(v+3/2)_a}{(1)_a\,(1/2)_a\,(v+2)_a}\left(\frac{x}{4}\right)^a a(a-1)\cdots(a-k+1)$$

$$= \frac{(-R)_k\,(R)_k\,(v+3/2)_k}{(1/2)_k\,(v+2)_k}\left(\frac{x}{4}\right)^k$$

$$\times\sum_{a=k}^{\infty}\frac{(-R+k)_{a-k}\,(R+k)_{a-k}\,(v+k+3/2)_{a-k}}{(1)_{a-k}\,(k+1/2)_{a-k}\,(v+k+2)_{a-k}}\left(\frac{x}{4}\right)^{a-k}$$

$$= \frac{(-R)_k\,(R)_k\,(v+3/2)_k}{(1/2)_k\,(v+2)_k}\left(\frac{x}{4}\right)^k {}_3F_2\left[\begin{matrix}-R+k,\,R+k,\,\frac{3}{2}+v+k\\[2pt]\frac{1}{2}+k,\,2+v+k\end{matrix};\frac{x}{4}\right].$$

Substituting this into (6.9) we obtain the first part of the following result.

---

[6]The hypergeometric function of parameters $a_1,\ldots,a_p$ and $b_1,\ldots,b_q$ is defined by

$$_pF_q\left[\begin{matrix}a_1,\ldots,a_p\\b_1,\ldots,b_q\end{matrix};z\right] = \sum_{k=0}^{\infty}\frac{(a_1)_k\cdots(a_p)_k}{k!\,(b_1)_k\cdots(b_q)_k}z^k,$$

where $(a)_0 := 1$ and $(a)_k := a(a+1)\cdots(a+k-1)$ for $k \geq 1$.

LEMMA 6.2. *We have*

$$T^{(n)}(R, v; x) = \frac{1}{R} \frac{(3/2)_v}{(2)_v}$$

$$\times \sum_{k=0}^{n} f_k \frac{(-R)_k (R)_k (v + 3/2)_k}{(1/2)_k (v + 2)_k} \left(\frac{x}{4}\right)^k {}_3F_2 \left[\begin{matrix} -R + k, \ R + k, \ \frac{3}{2} + v + k \ ; \frac{x}{4} \\ \frac{1}{2} + k, 2 + v + k \end{matrix}\right]$$

(6.10)

and

$$T'^{(n)}(R, v; x) = \frac{1}{R} \frac{(3/2)_v}{(1)_v}$$

$$\times \sum_{k=0}^{n} f_k \frac{(-R)_k (R)_k (v + 3/2)_k}{(3/2)_k (v + 1)_k} \left(\frac{x}{4}\right)^k {}_3F_2 \left[\begin{matrix} -R + k, \ R + k, \ \frac{3}{2} + v + k \ ; \frac{x}{4} \\ \frac{3}{2} + k, 1 + v + k \end{matrix}\right],$$

(6.11)

where the f_k's are as in (6.9) *(in particular $f_n = 1$).*

PROOF. Starting from (6.4), we obtain (6.11) by nearly the same calculation that proved (6.10). □

Since by Proposition 6.1 the boundary-influenced correlation ω_b is expressed in terms of the moments $M_{\alpha_1, \beta_1, \ldots, \alpha_m, \beta_m; \gamma_1, \delta_1, \ldots, \gamma_n, \delta_n}$ given by (6.6), which in turn depend on $T^{(n)}(R, v; x)$ and $T'^{(n)}(R, v; x)$, to determine the asymptotics of ω_b we need to understand the asymptotics of the $T^{(n)}$'s and $T'^{(n)}$'s. In the next section we deduce these latter two asymptotics from a result—stated in Proposition 7.2—whose technical proof we defer to Section 9. In Section 8 we show that the asymptotics of the $T^{(n)}$'s and $T'^{(n)}$'s found in Section 7 can be used to obtain the asymptotics of $M_{\alpha_1, \beta_1, \ldots, \alpha_m, \beta_m; \gamma_1, \delta_1, \ldots, \gamma_n, \delta_n}$.

7

The asymptotics of the $T^{(n)}$'s and $T'^{(n)}$'s

Given that in Theorem 2.1 the coordinates of the quadromers approach infinity as indicated by (2.3), we need more specifically to find the asymptotics of $T^{(n)}(R, qR + c; x)$ and $T'^{(n)}(R, qR + c; x)$, as $R \to \infty$. These are given by the following result.

PROPOSITION 7.1. *Let $q > 0$ be fixed rational number, and let $n \geq 0$ and c be fixed integers. Then for any real number $x \in (0, 1]$, we have*

$$\left| T^{(n)}(R, qR + c; x) - \frac{2}{\sqrt{\pi}} \frac{1}{\sqrt[4]{q^2 + \frac{x}{4-x}}} \frac{1}{R^{3/2}} \left(R\sqrt{\frac{x}{4-x}} \right)^n \right.$$

(7.1)

$$\left. \times \cos \left[R \arccos \left(1 - \frac{x}{2} \right) - \frac{1}{2} \arctan \frac{1}{q} \sqrt{\frac{x}{4-x}} + \frac{n\pi}{2} \right] \right| \leq M R^{n-5/2}$$

$$\left| T'^{(n)}(R, qR + c; x) - \frac{1}{\sqrt{\pi}} \frac{\sqrt[4]{q^2 + \frac{x}{4-x}}}{\sqrt{\frac{x}{4-x}}} \frac{1}{R^{3/2}} \left(R\sqrt{\frac{x}{4-x}} \right)^n \right.$$

(7.2)

$$\left. \times \cos \left[R \arccos \left(1 - \frac{x}{2} \right) + \frac{1}{2} \arctan \frac{1}{q} \sqrt{\frac{x}{4-x}} + \frac{(n-1)\pi}{2} \right] \right| \leq \frac{1}{\sqrt{x}} M' R^{n-5/2},$$

for $R \geq R_0$, where R_0, M and M' are independent of $x \in (0, 1]$.

In our proof of the above statements we make use of the following result, whose proof is presented in Section 9.

PROPOSITION 7.2. *Let $p(t)$ and $Q(t)$ be functions depending on the parameter $x \in [0, 1]$, defined on $(0, 1)$ by*

(7.3) $$p(t) = -q \ln t - i \arccos \left(1 - \frac{xt}{2} \right)$$

(7.4) $$Q(t) = \frac{t^l}{(1-t)^{1/2}} \frac{(4 - 2xt)^a}{(4 - xt)^b},$$

where $0 < q \in \mathbb{Q}$, $0 \leq a \in \mathbb{Z}$, $-1/2 \leq b \in \frac{1}{2}\mathbb{Z}$ and $l \in \frac{1}{2}\mathbb{Z}$ are all fixed. Then

(7.5) $$\left| \int_0^1 e^{-Rp(t)} Q(t) dt - F(R, x) \right| \leq M R^{-3/2},$$

for $R \geq R_0$, with R_0 and M independent of $x \in [0,1]$, and $F(R,x)$ given by

(7.6) $$F(R,x) = \frac{\sqrt{\pi}}{\sqrt{R}} \frac{(4-2x)^a/(4-x)^b}{\sqrt[4]{q^2 + \frac{x}{4-x}}} e^{i\left[R\arccos\left(1-\frac{x}{2}\right) - \frac{1}{2}\arctan\frac{1}{q}\sqrt{\frac{x}{4-x}}\right]}.$$

COROLLARY 7.3. *Under the assumptions of Proposition 7.2, we have*

(7.7) $$\left| \int_0^1 e^{-Rp(t)} Q(t)dt \right| \leq MR^{-1/2},$$

where M is independent of $x \in [0,1]$.

PROOF. The absolute value of the part of the right hand side of (7.6) not containing R can clearly be majorized, for $x \in [0,1]$, by a constant independent of x. The statement of the Corollary follows by combining this observation with (7.5). □

PROOF OF PROPOSITION 7.1. By [**24**, (10), p.58], a $_{p+1}F_{q+1}$ hypergeometric function can be written as an integral of a $_pF_q$ as

(7.8) $$_{p+1}F_{q+1}\left[\begin{matrix} \beta, \alpha_p \\ \beta+\sigma, \rho_q \end{matrix}; z\right] = \frac{\Gamma(\beta+\sigma)}{\Gamma(\beta)\Gamma(\sigma)} \int_0^1 t^{\beta-1}(1-t)^{\sigma-1}{}_pF_q\left[\begin{matrix} \alpha_p \\ \rho_q \end{matrix}; zt\right] dt,$$

provided $p \leq q+1$, $\mathrm{Re}\,\beta > 0$, $\mathrm{Re}\,\sigma > 0$, and $|z| < 1$ if $p = q+1$ (here α_p stands for a p-tuple, ρ_q for a q-tuple of parameters).

Taking $\beta = v+k+3/2$ and $\sigma = 1/2$, (7.8) yields

(7.9)
$$_3F_2\left[\begin{matrix} -R+k, R+k, \frac{3}{2}+v+k \\ \frac{1}{2}+k, 2+v+k \end{matrix}; \frac{x}{4}\right] = \frac{\Gamma(v+k+2)}{\Gamma(v+k+3/2)\Gamma(1/2)}$$
$$\times \int_0^1 t^{v+k+1/2}(1-t)^{-1/2}{}_2F_1\left[\begin{matrix} -R+k, R+k \\ \frac{1}{2}+k \end{matrix}; \frac{xt}{4}\right] dt.$$

On the other hand, from the definition of a $_2F_1$ hypergeometric function it readily follows that

$$\frac{d}{dz}{}_2F_1\left[\begin{matrix} a_1, a_2 \\ b \end{matrix}; z\right] = \frac{a_1 a_2}{b}{}_2F_1\left[\begin{matrix} a_1+1, a_2+1 \\ b+1 \end{matrix}; z\right].$$

Repeated application of this shows that

(7.10) $$_2F_1\left[\begin{matrix} -R+k, R+k \\ \frac{1}{2}+k \end{matrix}; \frac{xt}{4}\right] = \frac{(1/2)_k}{(-R)_k(R)_k(x/4)^k}\frac{d^k}{dt^k}{}_2F_1\left[\begin{matrix} -R, R \\ \frac{1}{2} \end{matrix}; \frac{xt}{4}\right].$$

However, the latter $_2F_1$ evaluates exactly (see for instance [**14**, p.1055, #1]):

(7.11) $$_2F_1\left[\begin{matrix} -R, R \\ \frac{1}{2} \end{matrix}; z\right] = \cos\left[R\arccos(1-2z)\right].$$

Replacing (7.9) and (7.10) into (6.10) and using (7.11) we obtain

$$
T^{(n)}(R, v; x) = \frac{1}{R} \sum_{k=0}^{n} f_k \frac{\frac{(-R)_k (R)_k (3/2)_{v+k}}{(1/2)_k (2)_{v+k}} \left(\frac{x}{4}\right)^k \frac{\Gamma(v+k+2)}{\Gamma(v+k+3/2)\Gamma(1/2)}}{\left(\frac{x}{4}\right)^k \frac{(-R)_k (R)_k}{(1/2)_k}}
$$

$$
\times \int_0^1 t^{v+k+1/2}(1-t)^{-1/2} \frac{d^k}{dt^k} \cos\left[R \arccos\left(1 - \frac{xt}{2}\right)\right] dt
$$

$$
(7.12) \qquad = \frac{2}{\pi R} \sum_{k=0}^{n} f_k \int_0^1 t^{v+k+1/2}(1-t)^{-1/2} \frac{d^k}{dt^k} \cos\left[R \arccos\left(1 - \frac{xt}{2}\right)\right] dt
$$

(for the second equality we also used $\Gamma(1/2) = \sqrt{\pi}$ and the recurrence $\Gamma(x+1) = x\Gamma(x)$).

By Lemma 7.4, the successive derivatives with respect to t of $\cos(R \arccos(1 - xt/2))$ can be written as

$$
\frac{d^k}{dt^k} \cos(R \arccos(1 - xt/2)) = \cos\left[R \arccos\left(1 - \frac{xt}{2}\right) + \frac{n\pi}{2}\right] \left(R\sqrt{\frac{x}{4t - xt^2}}\right)^k
$$

$$
(7.13) \qquad\qquad + O(R^{k-1}),
$$

where the terms of the omitted linear combination are of the form

$$
R^j x^s \cos(R \arccos(1 - xt/2))
$$

or

$$
R^j x^s \sin(R \arccos(1 - xt/2))
$$

times a function of type (7.4), with $0 \le j \le k-1$ and $s \ge 0$. (In fact, this is how the family of functions (7.4) was chosen, to contain all functions arising this way from successive derivatives of $\cos(R \arccos(1 - xt/2))$, and the analogous functions resulting when doing the same analysis to the $T'^{(n)}$'s.)

Since clearly $|x| \le 1$ throughout the range $x \in (0, 1]$, one sees by Corollary 7.3 that for $v = qR+c$, $q > 0$ and for any fixed $k \in \{0, \dots, n\}$, the total contribution of the lower order terms in (7.13) to the sum on the right hand side of (7.12) is bounded in absolute value by $L_k R^{k-3/2}$ for $R \ge R_k$, where R_k and L_k are independent of $x \in [0, 1]$, $k = 0, \dots, n$. Therefore, the combined contribution to $T^{(n)}(R, qR + c; x)$ of all lower order terms in (7.13), for $k = 0, \dots, n$, is bounded in absolute value by $LR^{k-5/2}$ for $R \ge \rho_0$, where ρ_0 and L are constants independent of $x \in [0, 1]$.

On the other hand, by using Corollary 7.3 as in the previous paragraph, one sees that the combined contribution of the leading terms of (7.13) for $k = 0, \dots, n-1$ to $T^{(n)}(R, qR + c; x)$ is again bounded in absolute value by $KR^{n-5/2}$ for all $R \ge \rho_1$ and $x \in (0, 1]$, for some constants K and ρ_1 independent of $x \in (0, 1]$.

Taking into account the leading term of (7.13) for $k = n$, we obtain by (7.12) and the previous two paragraphs that

$$
\left| T^{(n)}(R, qR + c; x) - \frac{2}{\pi R} \int_0^1 t^{qR} \frac{t^{n+c+1/2}}{(1-t)^{1/2}} \left(R\sqrt{\frac{x}{4t - xt^2}}\right)^n \right.
$$

$$
(7.14)
$$

$$
\left. \times \cos\left[R \arccos\left(1 - \frac{xt}{2}\right) + \frac{n\pi}{2}\right] dt \right| \le M_0 R^{n-5/2},
$$

for $R \geq \rho_2$, with ρ_2 and M_0 independent of $x \in [0,1]$. However, it is readily seen by Proposition 7.2 that

$$\left| \int_0^1 t^{qR} \frac{t^{n+c+1/2}}{(1-t)^{1/2}} \left(\sqrt{\frac{x}{4t-xt^2}} \right)^n \cos\left[R\arccos\left(1 - \frac{xt}{2}\right) + \frac{n\pi}{2} \right] dt \right.$$

$$- \frac{\sqrt{\pi}}{\sqrt{R}} \left(q^2 + \frac{x}{4-x} \right)^{-1/4} \left(\sqrt{\frac{x}{4-x}} \right)^n$$

$$\times \cos\left[R\arccos\left(1 - \frac{x}{2}\right) - \frac{1}{2}\arctan\frac{1}{q}\sqrt{\frac{x}{4-x}} + \frac{n\pi}{2} \right] \Bigg|$$

(7.15)
$$\leq M_1 R^{n-5/2},$$

for $R \geq \rho_3$, with ρ_3 and M_1 independent of $x \in [0,1]$. Relations (7.14) and (7.15) imply (7.1).

We now turn to proving (7.2). Because of the requirement $\operatorname{Re}\sigma > 0$ we cannot apply (7.8) directly to the $_3F_2$'s of (6.11), so we write first

$$_3F_2\left[\begin{matrix} -R+k,\, R+k,\, \frac{3}{2}+v+k \\ \frac{3}{2}+k,\, 1+v+k \end{matrix} ; \frac{x}{4} \right] =$$

$$\sum_{a \geq 0} \frac{(-R+k)_a\,(R+k)_a\,(1/2+v+k)_a \left(1 + \dfrac{a}{1/2+v+k}\right)}{a!\,(3/2+k)_a\,(1+v+k)_a} \left(\frac{x}{4}\right)^a$$

$$= {}_3F_2\left[\begin{matrix} -R+k,\, R+k,\, \frac{1}{2}+v+k \\ \frac{3}{2}+k,\, 1+v+k \end{matrix} ; \frac{x}{4} \right] +$$

$$\frac{1}{1/2+v+k} \frac{(-R+k)_1\,(R+k)_1\,(1/2+v+k)_1}{(3/2+k)_1\,(1+v+k)_1} \left(\frac{x}{4}\right)$$

$$\times \sum_{a \geq 1} \frac{(-R+k+1)_{a-1}\,(R+k+1)_{a-1}\,(3/2+v+k)_{a-1}}{(a-1)!\,(5/2+k)_{a-1}\,(2+v+k)_{a-1}} \left(\frac{x}{4}\right)^{a-1}$$

$$= {}_3F_2\left[\begin{matrix} -R+k,\, R+k,\, \frac{1}{2}+v+k \\ \frac{3}{2}+k,\, 1+v+k \end{matrix} ; \frac{x}{4} \right] +$$

(7.16)
$$\frac{x}{4} \frac{(-R+k)(R+k)}{(3/2+k)(1+v+k)}\, {}_3F_2\left[\begin{matrix} -R+k+1,\, R+k+1,\, \frac{3}{2}+v+k \\ \frac{5}{2}+k,\, 2+v+k \end{matrix} ; \frac{x}{4} \right].$$

The two $_3F_2$'s on the right hand side of (7.16) have the same form (the second is obtained from the first by replacing k by $k+1$). Equality (7.8) is applicable to them and yields

$$_3F_2\left[\begin{matrix} -R+k,\, R+k,\, \frac{1}{2}+v+k \\ \frac{3}{2}+k,\, 1+v+k \end{matrix} ; \frac{x}{4} \right] = \frac{\Gamma(v+k+1)}{\Gamma(v+k+1/2)\Gamma(1/2)}$$

(7.17)
$$\times \int_0^1 t^{v+k-1/2}(1-t)^{-1/2}\, {}_2F_1\left[\begin{matrix} -R+k,\, R+k \\ \frac{3}{2}+k \end{matrix} ; \frac{xt}{4} \right] dt.$$

Repeated application of the relation just before (7.10) shows that

$$(7.18) \quad {}_2F_1\left[\begin{matrix} -R+k,\, R+k \\ \frac{3}{2}+k \end{matrix};\frac{xt}{4}\right] = \frac{(3/2)_k}{(-R)_k\,(R)_k\,(x/4)^k}\frac{d^k}{dt^k}\,{}_2F_1\left[\begin{matrix} -R,\, R \\ \frac{3}{2} \end{matrix};\frac{xt}{4}\right].$$

To continue our analysis we need a closed formula for the ${}_2F_1$ on the right hand side of (7.18). We obtain it as follows. By [14, 8.962, #1] one has

$$P_n^{(\alpha,\beta)}(x) = \frac{(-1)^n\Gamma(n+1+\beta)}{n!\,\Gamma(1+\beta)}\,{}_2F_1\left[\begin{matrix} -n,\, n+\alpha+\beta+1 \\ 1+\beta \end{matrix};\frac{1+x}{2}\right],$$

where $P_n^{(\alpha,\beta)}(x)$ is the nth Jacobi polynomial of parameters α and β. Taking $\alpha = -3/2$, $\beta = 1/2$, this gives

$$(7.19) \quad {}_2F_1\left[\begin{matrix} -n,\, n \\ \frac{3}{2} \end{matrix};\frac{1+x}{2}\right] = \frac{n!\,\Gamma(3/2)}{(-1)^n\Gamma(n+3/2)}P_n^{(-\frac{3}{2},\frac{1}{2})}(x).$$

On the other hand, from [14, 8.962, #4] we obtain

$$(7.20) \quad \begin{aligned} P_n^{(\frac{1}{2},\frac{1}{2})}(x) &= \frac{\Gamma(2)\Gamma(n+3/2)}{\Gamma(n+2)\Gamma(3/2)}C_n^1(x) \\ &= \frac{\Gamma(2)}{\Gamma(3/2)}\frac{\Gamma(n+3/2)}{\Gamma(n+2)}\frac{\sin[(n+1)\arccos x]}{\sin(\arccos x)}, \end{aligned}$$

where C_n^λ is the nth ultraspherical polynomial of parameter λ, and at the last equality in (7.20) we used [14, 8.937, #1].

However, by [14, 8.961, #8] the Jacobi polynomials satisfy the recurrence

$$(2n+\alpha+\beta)P_n^{(\alpha-1,\beta)}(x) = (n+\alpha+\beta)P_n^{(\alpha,\beta)}(x) - (n+\beta)P_{n-1}^{(\alpha,\beta)}(x).$$

By this the explicit formula (7.20) for $P_n^{(\frac{1}{2},\frac{1}{2})}(x)$ yields one for $P_n^{(-\frac{1}{2},\frac{1}{2})}(x)$, which in turn, by another application of the above recurrence, yields an explicit formula for $P_n^{(-\frac{3}{2},\frac{1}{2})}(x)$. Substituting this into (7.19) one obtains after simplifications that

$$(7.21) \quad \begin{aligned} {}_2F_1\left[\begin{matrix} -n,\, n \\ \frac{3}{2} \end{matrix};\frac{t}{4}\right] &= \frac{2n}{4n^2-1}\sqrt{\frac{4-t}{t}}\sin\left[n\arccos\left(1-\frac{t}{2}\right)\right] \\ &\quad - \frac{1}{4n^2-1}\cos\left[n\arccos\left(1-\frac{t}{2}\right)\right]. \end{aligned}$$

Expressing the $_3F_2$'s in (6.11) with the use of (7.17), (7.18) and (7.21), (6.11) becomes

$$T'^{(n)}(R, v; x) =$$

$$\frac{1}{R} \sum_{k=0}^{n} f_k \frac{\frac{(-R)_k (R)_k (3/2)_{v+k}}{(3/2)_k (1)_{v+k}} \left(\frac{x}{4}\right)^k \frac{\Gamma(v+k+1)}{\Gamma(v+k+1/2)\Gamma(1/2)}}{\frac{(-R)_k (R)_k}{(3/2)_k} \left(\frac{x}{4}\right)^k}$$

$$\times \int_0^1 t^{v+k-1/2}(1-t)^{-1/2} \frac{d^k}{dt^k} \left\{ \frac{2R}{4R^2-1} \sqrt{\frac{4-xt}{xt}} \sin\left[R\arccos\left(1-\frac{xt}{2}\right)\right] \right.$$

$$\left. - \frac{1}{4R^2-1} \cos\left[R\arccos\left(1-\frac{xt}{2}\right)\right] \right\} dt$$

$$+ \frac{1}{R} \sum_{k=0}^{n} f_k \frac{\frac{(-R)_k (R)_k (3/2)_{v+k}}{(3/2)_k (1)_{v+k}} \left(\frac{x}{4}\right)^k \frac{\Gamma(v+k+2)}{\Gamma(v+k+3/2)\Gamma(1/2)}}{\frac{(-R)_{k+1} (R)_{k+1}}{(3/2)_{k+1}} \left(\frac{x}{4}\right)^{k+1}} \frac{x}{4} \frac{(-R+k)(R+k)}{(3/2+k)(1+v+k)}$$

$$\times \int_0^1 t^{v+k+1/2}(1-t)^{-1/2} \frac{d^{k+1}}{dt^{k+1}} \left\{ \frac{2R}{4R^2-1} \sqrt{\frac{4-xt}{xt}} \sin\left[R\arccos\left(1-\frac{xt}{2}\right)\right] \right.$$

$$\left. - \frac{1}{4R^2-1} \cos\left[R\arccos\left(1-\frac{xt}{2}\right)\right] \right\} dt,$$

which when substituting $v = qR + c$ becomes after simplifications

$$T'^{(n)}(R, qR+c; x) = \frac{4}{\pi(4R^2-1)} \sum_{k=0}^{n} f_k(qR+c+k+1/2) \int_0^1 t^{qR} \frac{t^{k+c-1/2}}{(1-t)^{1/2}}$$

$$\times \frac{d^k}{dt^k} \left\{ \sqrt{\frac{4-xt}{xt}} \sin\left[R\arccos\left(1-\frac{xt}{2}\right)\right] - \frac{1}{2R} \cos\left[R\arccos\left(1-\frac{xt}{2}\right)\right] \right\} dt$$

$$+ \frac{4}{\pi} \frac{1}{4R^2-1} \sum_{k=0}^{n} f_k \int_0^1 t^{qR} \frac{t^{k+c+1/2}}{(1-t)^{1/2}}$$

$$\times \frac{d^{k+1}}{dt^{k+1}} \left\{ \sqrt{\frac{4-xt}{xt}} \sin\left[R\arccos\left(1-\frac{xt}{2}\right)\right] - \frac{1}{2R} \cos\left[R\arccos\left(1-\frac{xt}{2}\right)\right] \right\} dt.$$

(7.22)

Let $F(t) = \sqrt{(4-xt)/(xt)} \sin(R\arccos(1-xt/2)) - 1/(2R) \cos(R\arccos(1-xt/2))$. Lemma 7.4 implies that

$$F^{(k)}(t) = \sqrt{\frac{4-xt}{xt}} \cos\left[R\arccos\left(1-\frac{xt}{2}\right) + \frac{(n-1)\pi}{2}\right] \left(R\sqrt{\frac{x}{4t-xt^2}}\right)^k$$

(7.23)

$$+ O(R^{k-1}),$$

where each term of the omitted linear combination is $R^j x^s \cos(R\arccos(1-xt/2))$ or $R^j x^s \sin(R\arccos(1-xt/2))$ times a function of type (7.4), for some $j \leq k-1$ and $s \geq -1/2$.

Since there are only a finite number of omitted terms in (7.23) for each $k = 0, \ldots, n$, we obtain by (7.22) and Corollary 7.3 that their contribution to the first sum in (7.22) is bounded in absolute value by $L_k' R^{k-3/2}/\sqrt{x}$ for all $R \geq R_k'$ and $x \in (0,1]$, where L_k' and R_k', $k = 0, \ldots, n$, are some constants independent of $x \in (0,1]$. Therefore, the overall contribution of the non-leading terms in (7.23) to the first of the two terms on the right hand side of (7.22) is bounded in absolute value by $L_1' R^{n-5/2}/\sqrt{x}$ for all $R \geq \rho'$ and $x \in (0,1]$, where L_1' and ρ' are constants independent of $x \in (0,1]$. Similarly, the overall contribution of the non-leading terms in (7.23) to the second term on the right hand side of (7.22) is seen to be bounded in absolute value by $L_2' R^{n-5/2}/\sqrt{x}$ for all $R \geq \rho''$ and $x \in (0,1]$, with constants L_2' and ρ'' independent of $x \in (0,1]$. Therefore the total contribution of the non-leading terms in (7.23) to $T'^{(n)}(R, qR + c; x)$ is bounded in absolute value by $L' R^{n-5/2}/\sqrt{x}$ for all $R \geq \rho_0'$ and $x \in (0,1]$, for some constants L' and ρ_0' independent of $x \in (0,1]$.

On the other hand, by using Corollary 7.3 as in the previous paragraph, one sees that the combined contribution of the leading terms of (7.23) for $k = 0, \ldots, n - 1$ to $T'^{(n)}(R, qR + c; x)$ is again bounded in absolute value by $K' R^{n-5/2}/\sqrt{x}$ for all $R \geq \rho_1'$ and $x \in (0,1]$, for some constants K' and ρ_1' independent of $x \in (0,1]$.

By the previous two paragraphs and (7.23) we obtain from the $k = n$ terms of (7.22) that

$$
\begin{aligned}
\Bigg| T'^{(n)}(R, qR + c; x) - \Bigg\{ & \frac{q}{\pi R} \int_0^1 t^{qR} \frac{t^{c-1/2}}{(1-t)^{1/2}} \sqrt{\frac{4 - xt}{xt}} \left(R\sqrt{\frac{xt}{4 - xt}} \right)^n \\
& \times \cos\left[R \arccos\left(1 - \frac{xt}{2} \right) + \frac{(n-1)\pi}{2} \right] dt \\
+ & \frac{1}{\pi R^2} \int_0^1 t^{qR} \frac{t^{c-1/2}}{(1-t)^{1/2}} \sqrt{\frac{4 - xt}{xt}} \left(R\sqrt{\frac{xt}{4 - xt}} \right)^{n+1}
\end{aligned}
$$

(7.24)

$$
\times \cos\left[R \arccos\left(1 - \frac{xt}{2} \right) + \frac{n\pi}{2} \right] dt \Bigg\} \Bigg| \leq \frac{1}{\sqrt{x}} M_0' R^{n-5/2},
$$

for all $R \geq \rho_2'$ and $x \in (0,1]$, where the constants M_0' and ρ_2' are independent of $x \in (0,1]$.

Denote by I_1 and I_2 the first and second integrals on the right hand side of (7.24), respectively. They clearly satisfy the hypothesis of Proposition 7.2. By (7.5) and (7.6) we obtain

$$
\begin{aligned}
\Bigg| \frac{q}{\pi R} I_1 - \mathrm{Re} \Bigg\{ & \frac{q}{\sqrt{\pi R}} \left(q^2 + \frac{x}{4 - x} \right)^{-1/4} \left(R\sqrt{\frac{x}{4 - x}} \right)^{n-1} \\
& \times \exp\left(i\left[R \arccos\left(1 - \frac{x}{2} \right) - \frac{1}{2}\arctan\frac{1}{q}\sqrt{\frac{x}{4 - x}} + \frac{(n-1)\pi}{2} \right] \right) dt \Bigg\} \Bigg|
\end{aligned}
$$

(7.25)

$$
\leq M_1' R^{n-5/2} x^{(n-1)/2}
$$

and

$$\left| \frac{1}{\pi R^2} I_2 - \mathrm{Re} \left\{ \frac{1}{\sqrt{\pi} R^{3/2}} \left(q^2 + \frac{x}{4-x} \right)^{-1/4} \left(R\sqrt{\frac{x}{4-x}} \right)^n \right. \right.$$

$$\left. \left. \times \exp \left(i \left[R \arccos \left(1 - \frac{x}{2} \right) - \frac{1}{2} \arctan \frac{1}{q} \sqrt{\frac{x}{4-x}} + \frac{n\pi}{2} \right] \right) dt \right\} \right|$$

(7.26)
$$\leq M_2' R^{n-5/2} x^{n/2},$$

for $R \geq \rho_3'$, where ρ_3', M_1' and M_2' are constants independent of $x \in (0,1]$.

Using the fact that

$$q + i\sqrt{\frac{x}{4-x}} = \sqrt{q^2 + \frac{x}{4-x}} \exp \left(i \arctan \frac{1}{q} \sqrt{\frac{x}{4-x}} \right),$$

the two second terms on the left hand sides of (7.25) and (7.26) are readily seen to add up precisely to the second term on the left hand side of (7.2). Since in (7.25) and (7.26) the integer n is nonnegative and since $|x| \leq 1$, inequalities (7.25), (7.26) and (7.24) imply (7.2). $\qquad \square$

LEMMA 7.4. *Let the functions* $F_1(t)$, $F_2(t)$ *and* $F_3(t)$ *be given by*

$$F_1(t) = \cos \left[R \arccos \left(1 - \frac{xt}{2} \right) \right]$$

$$F_2(t) = \sin \left[R \arccos \left(1 - \frac{xt}{2} \right) \right]$$

$$F_3(t) = \sqrt{\frac{4-xt}{xt}} \sin \left[R \arccos \left(1 - \frac{xt}{2} \right) \right],$$

where $x \in (0,1]$ *is a parameter. Then for any* $n \geq 0$ *one has*

(7.27)
$$F_1^{(n)}(t) = \left(R\sqrt{\frac{x}{4t-xt^2}} \right)^n \cos \left[R \arccos \left(1 - \frac{xt}{2} \right) + \frac{n\pi}{2} \right] + \sum_{\nu \in V_1} c_\nu^{(1)} R^{j_\nu^{(1)}} x^{l_\nu^{(1)}} Q_\nu^{(1)}$$

(7.28)
$$F_2^{(n)}(t) = \left(R\sqrt{\frac{x}{4t-xt^2}} \right)^n \sin \left[R \arccos \left(1 - \frac{xt}{2} \right) + \frac{n\pi}{2} \right] + \sum_{\nu \in V_2} c_\nu^{(2)} R^{j_\nu^{(2)}} x^{l_\nu^{(2)}} Q_\nu^{(2)}$$

$$F_3^{(n)}(t) = \sqrt{\frac{4-xt}{xt}} \left(R\sqrt{\frac{x}{4t-xt^2}} \right)^n \cos \left[R \arccos \left(1 - \frac{xt}{2} \right) + \frac{(n-1)\pi}{2} \right]$$

(7.29)
$$+ \sum_{\nu \in V_3} c_\nu^{(3)} R^{j_\nu^{(3)}} x^{l_\nu^{(3)}} Q_\nu^{(3)},$$

where $l_\nu^{(1)}, l_\nu^{(2)} \geq 0$, $l_\nu^{(3)} \geq -1/2$, *for all* ν, *and for* $k = 1,2,3$ *and all* $\nu \in V_k$ *one has*

(i) V_k *is some finite set*
(ii) $c_\nu^{(k)}, j_\nu^{(k)}, l_\nu^{(k)} \in \mathbb{Q}$
(iii) $j_\nu^{(k)} \leq n-1$

(iv) $Q_\nu^{(k)}$ *is a function of type* (7.4) *multiplied by either* $\cos(R\arccos(1-xt/2))$ *or* $\sin(R\arccos(1-xt/2))$.

PROOF. It readily follows by induction on n that for $n \geq 1$ one has

$$
(7.30) \qquad \frac{d^n}{dt^n} e^{h(t)} = e^{h(t)} \sum_{k=1}^{n} \sum_{\substack{i_1,\ldots,i_k \geq 1 \\ i_1+\cdots+i_k=n}} \alpha_I h^{(i_1)}(t)\cdots h^{(i_k)}(t),
$$

with coefficients $\alpha_I = \alpha_{(i_1,\ldots,i_k)} \in \mathbb{Z}$ and $\alpha_{(1,\ldots,1)} = 1$.

A similar argument shows that for $n \geq 1$ one has

$$
(7.31) \qquad \frac{d^n}{dt^n}[g(t)]^{-1/2} = [g(t)]^{-1/2} \sum_{k=1}^{n} \sum_{i_0,i_1,\ldots,i_k \geq 1} \beta_I [g(t)]^{-i_0} g^{(i_1)}(t)\cdots g^{(i_k)}(t),
$$

where only finitely many of the coefficients $\beta_I = \beta_{(i_0,i_1,\ldots,i_k)} \in \mathbb{Q}$ are nonzero.

Choosing $h(t) = iR\arccos(1-xt/2)$, the derivatives on the left hand side of (7.27) can be expressed as

$$
(7.32) \qquad F_1^{(n)}(t) = \operatorname{Re} \frac{d^n}{dt^n} e^{h(t)}.
$$

For our choice of h we obtain

$$
(7.33) \qquad h'(t) = iR\sqrt{\frac{x}{4t-xt^2}} = iR\sqrt{x}[g(t)]^{-1/2},
$$

with $g(t) = 4t - xt^2$. Since $g'(t) = 4 - 2xt$, $g''(t) = -2x$, and the higher derivatives of g are zero, (7.33) and (7.31) imply that for $j \geq 1$ we have

$$
(7.34) \qquad \frac{d^j}{dt^j} h'(t) = iR\sqrt{x} \sum_\nu \beta_\nu \frac{(4-2xt)^{a_\nu}(-2x)^{s_\nu}}{(4t-xt^2)^{1/2+b_\nu}},
$$

where the sum is finite, $\beta_\nu \in \mathbb{Q}$ and $a_\nu, b_\nu, s_\nu \geq 0$ are integers. From the expression of $h'(t)$ it is apparent that (7.34) holds also for $j = 0$.

Substituting (7.34) in (7.30) we obtain for $n \geq 1$ that

$$
\frac{d^n}{dt^n} e^{h(t)} = e^{h(t)}[h'(t)]^n + e^{h(t)} \sum_{k=1}^{n-1} \sum_{\substack{i_1,\ldots,i_k \geq 1 \\ i_1+\cdots+i_k=n}} \alpha_I h^{(i_1)}(t)\cdots h^{(i_1)}(t)
$$

$$
(7.35)
$$

$$
= e^{h(t)}\left(iR\sqrt{\frac{x}{4t-xt^2}}\right)^n + e^{h(t)} \sum_\nu \gamma_\nu (iR\sqrt{x})^{n_\nu} \frac{(4-2xt)^{a_\nu} x^{s_\nu}}{(4t-xt^2)^{b_\nu}},
$$

where the sum is finite, $n-1 \geq n_\nu \in \mathbb{Z}$, $0 \leq a_\nu, s_\nu \in \mathbb{Z}$, $0 \leq b_\nu \in \frac{1}{2}\mathbb{Z}$, and $\gamma_\nu \in \mathbb{Q}$. Relation (7.35) clearly holds also for $n = 0$. By (7.32), taking the real and imaginary parts in (7.35) one obtains (7.27) and (7.28), respectively.

To prove (7.29), apply Leibniz's formula for the derivatives of the product $f_1(t)f_2(t)$ defining $F_3(t)$, with $f_1(t) = \sqrt{(4-xt)/(xt)}$, $f_2(t) = \sin(R(\arccos(1-xt/2))$. It is clear by (7.28) that the highest power of R in the resulting terms is n, and it occurs in only one of these terms, namely in $f_1(t)$ times the highest order term in $f_2^{(n)}(t)$. By (7.28) it also follows that the leading term in R of $F_3^{(n)}(t)$ is the first term on the right hand side of (7.29).

Furthermore, (7.28) implies that the successive derivatives of the second factor $f_2(t)$ have the form of the summand on the right hand side of (7.29). Therefore, to finish the proof it is enough to show that the derivatives of $f_1(t)$ are also of this form.

We have

$$(7.36) \qquad \frac{d^n}{dt^n} \sqrt{\frac{4 - xt}{xt}} = x^{-1/2} \frac{d^n}{dt^n} \sqrt{\frac{4}{t} - x}.$$

One easily obtains by induction on n that for $n \geq 1$

$$(7.37) \qquad \frac{d^n}{dt^n} [g(t)]^{1/2} = [g(t)]^{1/2} \sum_{k=1}^{n} \sum_{i_0, i_1, \ldots, i_k \geq 1} \delta_I [g(t)]^{-i_0} g^{(i_1)}(t) \cdots g^{(i_k)}(t),$$

where the sum is finite and the coefficients $\delta_I = \beta_{(i_0, i_1, \ldots, i_k)} \in \mathbb{Q}$.

Choose $g(t) = \frac{4}{t} - x$. Since $g^{(j)}(t) = 4(-1)^j j! \, t^{-j-1}$, $j \geq 1$, it follows by (7.37) and (7.36) that for $n \geq 1$

$$(7.38) \qquad \frac{d^n}{dt^n} \sqrt{\frac{4 - xt}{xt}} = x^{-1/2} \sqrt{\frac{4}{t} - x} \sum_{k=1}^{n} \sum_{\nu} \delta_\nu \left(\frac{4}{t} - x \right)^{-s_\nu} t^{l_\nu},$$

where $0 \leq s_\nu, l_\nu \in \mathbb{Z}$. Since the terms in (7.38) have the form of the summand on the right hand side of (7.29), and since this is also true of the left hand side of (7.38) for $n = 0$, the proof of (7.29) is complete. $\qquad \square$

Replacement of the $T^{(k)}$'s and $T'^{(k)}$'s by their asymptotics does not affect the asymptotics of the moments M

Denote the approximants of the $T^{(k)}$'s and $T'^{(k)}$'s in Proposition 7.1 by

$$F_k(R, q; x) = \frac{2}{\sqrt{\pi}} \frac{1}{\left(q^2 + \frac{x}{4-x}\right)^{1/4}} \frac{1}{R^{3/2}} \left(R\sqrt{\frac{x}{4-x}}\right)^k$$

(8.1)

$$\times \cos\left[R \arccos\left(1 - \frac{x}{2}\right) - \frac{1}{2}\arctan\frac{1}{q}\sqrt{\frac{x}{4-x}} + \frac{k\pi}{2}\right]$$

and

$$F'_k(R, q; x) = \frac{1}{\sqrt{\pi}} \frac{\left(q^2 + \frac{x}{4-x}\right)^{1/4}}{\sqrt{\frac{x}{4-x}}} \frac{1}{R^{3/2}} \left(R\sqrt{\frac{x}{4-x}}\right)^k$$

(8.2)

$$\times \cos\left[R \arccos\left(1 - \frac{x}{2}\right) + \frac{1}{2}\arctan\frac{1}{q}\sqrt{\frac{x}{4-x}} + \frac{(k-1)\pi}{2}\right].$$

PROPOSITION 8.1. *Consider the moments $M_{\alpha_1,\beta_1,\ldots,\alpha_m,\beta_m;\gamma_1,\delta_1,\ldots,\gamma_n,\delta_n}$ defined by (6.6). As the variables R_i, v_i, R'_j, v'_j, $i = 1,\ldots,m$, $j = 1,\ldots,n$ approach infinity as specified by (2.3), we have*

$$M_{\alpha_1,\beta_1,\ldots,\alpha_m,\beta_m;\gamma_1,\delta_1,\ldots,\gamma_n,\delta_n} = \int_0^1 \cdots \int_0^1 \prod_{i=1}^m \prod_{j=1}^n (x_{ij} y_{ij} z_{ij} w_{ij})^{q_i R_i + q'_j R'_j + c_i + c'_j + 1}$$

$$\times F_{\alpha_1}(R_1, q_1; \prod_{j=1}^n x_{1j} y_{1j})\, F_{\beta_1}(R_1, q_1; \prod_{j=1}^n z_{1j} w_{1j}) \cdots$$

$$\cdots F_{\alpha_m}(R_m, q_m; \prod_{j=1}^n x_{mj} y_{mj})\, F_{\beta_m}(R_m, q_m; \prod_{j=1}^n z_{mj} w_{mj})$$

$$\times F'_{\gamma_1}(R'_1, q'_1; \prod_{i=1}^m x_{i1} z_{i1})\, F'_{\delta_1}(R'_1, q'_1; \prod_{i=1}^m y_{i1} w_{i1}) \cdots$$

$$\cdots F'_{\gamma_n}(R'_n, q'_n; \prod_{i=1}^m x_{in} z_{in})\, F'_{\delta_n}(R'_n, q'_n; \prod_{i=1}^m y_{in} w_{in})\, dx_{11} \cdots dw_{mn}$$

(8.3)
$$+ O\left(R^{-4mn + \sum_{i=1}^m (\alpha_i + \beta_i) + \sum_{i=1}^n (\gamma_i + \delta_i) - 3m - 3n - 1} \right),$$

where the integration variables are x_{ij}, y_{ij}, z_{ij}, w_{ij}, $i = 1, \ldots, m$, $j = 1, \ldots, n$.

PROOF. By Proposition 7.1, for any fixed $k \geq 0$ we have

(8.4)
$$|T^{(k)}(R, qR + c; x) - F_k(R, q; x)| \leq M R^{k-5/2}$$

(8.5)
$$|T'^{(k)}(R, qR + c; x) - F'_k(R, q; x)| \leq x^{-1/2} M' R^{k-5/2}$$

for $R \geq R_0$, where the constants R_0, M and M' are independent of $x \in (0, 1]$.

From (8.1) and (8.2) it is clearly seen that there exist constants M_1 and M'_1 so that

(8.6)
$$|F_k(R, q; x)| \leq M_1 R^{k-3/2}$$

(8.7)
$$|F'_k(R, q; x)| \leq x^{-1/2} M'_1 R^{k-3/2},$$

for all $k \geq 0$, $R \geq 0$ and $x \in (0, 1]$.

By (8.4)–(8.7) it follows that for any fixed $k \geq 0$

(8.8)
$$|T^{(k)}(R, qR + c; x)| \leq M_2 R^{k-3/2}$$

(8.9)
$$|T'^{(k)}(R, qR + c; x)| \leq x^{-1/2} M'_2 R^{k-3/2},$$

for $R \geq R'_0$, for some constants R'_0, M_2 and M'_2 that are independent of $x \in (0, 1]$.

By (6.6) we have

$$
\left| M_{\alpha_1,\beta_1,\ldots,\alpha_m,\beta_m;\gamma_1,\delta_1,\ldots,\gamma_n,\delta_n} - \int_0^1 \cdots \int_0^1 \prod_{i=1}^m \prod_{j=1}^n (x_{ij}y_{ij}z_{ij}w_{ij})^{q_i R_i + q_j' R' j + c_i + c_j' + 1} \right.
$$

$$
\times F_{\alpha_1}(R_1,q_1;\prod_{j=1}^n x_{1j}y_{1j})\, F_{\beta_1}(R_1,q_1;\prod_{j=1}^n z_{1j}w_{1j}) \cdots
$$

$$
\cdots F_{\alpha_m}(R_m,q_m;\prod_{j=1}^n x_{mj}y_{mj})\, F_{\beta_m}(R_m,q_m;\prod_{j=1}^n z_{mj}w_{mj})
$$

$$
\times F'_{\gamma_1}(R_1',q_1';\prod_{i=1}^m x_{i1}z_{i1})\, F'_{\delta_1}(R_1',q_1';\prod_{i=1}^m y_{i1}w_{i1}) \cdots
$$

$$
\left. \cdots F'_{\gamma_n}(R_n',q_n';\prod_{i=1}^m x_{in}z_{in})\, F'_{\delta_n}(R_n',q_n';\prod_{i=1}^m y_{in}w_{in})\, dx_{11}\cdots dw_{mn} \right|
$$

$$
= \left| \int_0^1 \cdots \int_0^1 \prod_{i=1}^m \prod_{j=1}^n (x_{ij}y_{ij}z_{ij}w_{ij})^{q_i R_i + q' j R_j' + c_i + c_j' + 1} \right.
$$

$$
\times \{ T^{(\alpha_1)}(R_1,v_1;\prod_{j=1}^n x_{1j}y_{1j})\, T^{(\beta_1)}(R_1,v_1;\prod_{j=1}^n z_{1j}w_{1j}) \cdots
$$

$$
\cdots T^{(\alpha_m)}(R_m,v_m;\prod_{j=1}^n x_{mj}y_{mj})\, T^{(\beta_m)}(R_m,v_m;\prod_{j=1}^n z_{mj}w_{mj})
$$

$$
\times T'^{(\gamma_1)}(R_1',v_1';\prod_{i=1}^m x_{i1}z_{i1})\, T'^{(\delta_1)}(R_1',v_1';\prod_{i=1}^m y_{i1}w_{i1}) \cdots
$$

$$
\cdots T'^{(\gamma_n)}(R_n',v_n';\prod_{i=1}^m x_{in}z_{in})\, T'^{(\delta_n)}(R_n',v_n';\prod_{i=1}^m y_{in}w_{in})
$$

$$
- F_{\alpha_1}(R_1,q_1;\prod_{j=1}^n x_{1j}y_{1j})\, F_{\beta_1}(R_1,q_1;\prod_{j=1}^n z_{1j}w_{1j}) \cdots
$$

$$
\cdots F_{\alpha_m}(R_m,q_m;\prod_{j=1}^n x_{mj}y_{mj})\, F_{\beta_m}(R_m,q_m;\prod_{j=1}^n z_{mj}w_{mj})
$$

$$
\times F'_{\gamma_1}(R_1',q_1';\prod_{i=1}^m x_{i1}z_{i1})\, F'_{\delta_1}(R_1',q_1';\prod_{i=1}^m y_{i1}w_{i1}) \cdots
$$

$$
\left. \cdots F'_{\gamma_n}(R_n',q_n';\prod_{i=1}^m x_{in}z_{in})\, F'_{\delta_n}(R_n',q_n';\prod_{i=1}^m y_{in}w_{in}) \}\, dx_{11}\cdots dw_{mn} \right|.
$$

(8.10)

Clearly, for any $2l$ quantities f_i, g_i, $i = 1, \ldots, l$ one has

$$f_1 \cdots f_l - g_1 \cdots g_l = f_1 \cdots f_{l-1}(f_l - g_l) + f_1 \cdots f_{l-2}(f_{l-1} - g_{l-1})g_l$$

(8.11)

$$+ f_1 \cdots f_{l-3}(f_{l-2} - g_{l-2})g_{l-1}g_l + \cdots + (f_1 - g_1)g_2 \cdots g_l.$$

Apply (8.11) to the expression E in the curly braces on the right hand side of (8.10): we have $l = 2m + 2n$, the f_i's become $T^{(k)}$'s or $T'^{(k)}$'s, and the g_i's become F_k's or F'_k's.

This results in expressing E as a sum of $2m + 2n$ products of the form $h_1 \cdots h_{2m+2n}$, where exactly one of the h_i's, say h_{i_0}, is a difference $T^{(k)} - F_k$ or $T'^{(k)} - F'_k$, all others being of the form $T^{(k)}$, $T'^{(k)}$, F_k or F'_k.

Furthermore, exactly $2n$ of the h_i's are of the form $T'^{(k)}$, F'_k, or $T'^{(k)} - F'_k$, and the set of their x-arguments is always

(8.12)
$$\left\{ \prod_{i=1}^{m} x_{i1}z_{i1}, \prod_{i=1}^{m} y_{i1}w_{i1}, \ldots, \prod_{i=1}^{m} x_{in}z_{in}, \prod_{i=1}^{m} y_{in}w_{in} \right\}.$$

Since the $2n$ subsets of variables that are multiplied together in the elements of (8.12) form a partition of the set of our $4mn$ variables $x_{ij}, y_{ij}, z_{ij}, w_{ij}, i = 1, \ldots, m$, $j = 1, \ldots, n$, inequalities (8.4)–(8.9), and the fact that $|h_{i_0}|$ can be bounded using the sharper (8.4) or (8.5), imply

$$|h_1 \cdots h_{2m+2n}| \leq \bar{M}^{2m+2n} R^{\sum_{i=1}^{m}(\alpha_i + \beta_i) + \sum_{i=1}^{n}(\gamma_i + \delta_i) - 3m - 3n - 1}$$

(8.13)

$$\times \prod_{i=1}^{m} \prod_{j=1}^{n} (x_{ij}y_{ij}z_{ij}w_{ij})^{-1/2},$$

for all $R \geq R_0''$, where the constants R_0'' and $\bar{M}$ are independent of $x \in (0, 1]$. Taking into account also the contribution of the double product in (8.10) to the integrand of (8.10), by (8.13) the absolute value of the latter is majorized by

$$(2m + 2n)\bar{M}^{2m+2n} R^{\sum_{i=1}^{m}(\alpha_i + \beta_i) + \sum_{i=1}^{n}(\gamma_i + \delta_i) - 3m - 3n - 1}$$

$$\times \prod_{i=1}^{m} \prod_{j=1}^{n} (x_{ij}y_{ij}z_{ij}w_{ij})^{q_i R_i + q'_j R'_j + c_i + c'_j + 1/2}.$$

By assumptions (2.3), this implies that (8.3) holds. $\qquad \square$

Proof of Proposition 7.2

For fixed $x \in [0, 1]$, the asymptotics of the integral in (7.5) can be readily obtained by Laplace's method for contour integrals, as it is presented for instance in [**28**, Ch.4, §6.1]. However, what (7.5) states is the existence of a *uniform error bound* for the Laplace approximation, for $x \in [0, 1]$. We obtain this by extending the arguments in the proof of [**28**, Theorem 6.1, p. 125] to the case when the integrand depends on a parameter.

For $t \in (0, 1)$, write the factor $(1 - t)^{1/2}$ in the denominator of $Q(t)$ in (7.4) as $(1 - t)^{1/2} = -i(t - 1)^{1/2}$, where the square root on the right hand side has its principal determination. The integral in (7.5) becomes

$$(9.1) \qquad I(R) = -i \int_1^0 e^{-Rp(t)} \tilde{Q}(t) dt =: -i\tilde{I}(R),$$

where $p(t)$ is given by (7.3) and

$$(9.2) \qquad \tilde{Q}(t) = \frac{t^l}{(t - 1)^{1/2}} \frac{(4 - 2xt)^a}{(4 - xt)^b}, \quad t \in (0, 1),$$

where as in (7.4) $0 \le a \in \mathbb{Z}$, $-1/2 \le b \in \frac{1}{2}\mathbb{Z}$, $l \in \frac{1}{2}\mathbb{Z}$, and the square root on the right hand side of (9.2) has the principal determination.

Regard the integral $\tilde{I}(R)$ defined by (9.1) as a contour integral over the path $\mathcal{P} = [1, 0]$ (a line segment) in the complex plane. Choose the principal determinations for all the multiple-valued maps in the expressions (7.3) and (9.2) defining $p(t)$ and $\tilde{Q}(\mathcal{P})$. Note that for any fixed $x \in [0, 1]$ the integral $\tilde{I}(R)$ satisfies the following properties:

(*i*) $p(t)$ and $\tilde{Q}(t)$ are independent of R, single-valued and holomorphic in the pointed open disk $\dot{D}(1, 1) = D(1, 1) \setminus \{1\}$.

(*ii*) $\mathcal{P}$ is independent of R, and $\mathcal{P}_{(1,0)}$ (i.e., the path $\mathcal{P}$ less its endpoints) is contained in $\dot{D}(1, 1)$.

(*iii*) For $t \in D(1, 1)$, the functions $p(t)$ and $\tilde{Q}(t)$ can be expanded in convergent series as

$$p(t) = p(1) + \sum_{s=0}^{\infty} p_s(t - 1)^{s+1},$$

where $p(1) = -i\arccos(1 - x/2)$ and $p_0 = -q - i\sqrt{x/(4 - x)}$, and

$$\tilde{Q}(t) = \sum_{s=0}^{\infty} q_s(t - 1)^{s-1/2},$$

where $q_0 = (4 - 2x)^a/(4 - x)^b$ and $(t - 1)^{1/2}$ has its principal determination.

(iv) $\tilde{I}(R)$ converges at 0 absolutely and uniformly with respect to $R \geq -l/q$ and $x \in [0,1]$.

(v) $\mathrm{Re}\{p(t) - p(1)\}$ is positive when $t \in (0,1)$, and is bounded away from 0 uniformly with respect to $x \in [0,1]$ as $t \to 0$ along $\mathcal{P}$.

To avoid interruption in proving Proposition 7.2, we phrase three facts we need in the proof as Lemmas 9.1–9.3, and include them at the end of this section.

Consider the map $t \mapsto v(t)$, $|t - 1| < 1$, given by

$$(9.3) \qquad v(t) = p(t) - p(1) = -q \ln t - i \arccos\left(1 - \frac{xt}{2}\right) + i \arccos\left(1 - \frac{x}{2}\right).$$

Let U_x be a neighborhood of $t = 1$ and D a disk centered at $v = 0$ satisfying the statement of Lemma 9.1 (in particular, D is independent of $x \in [0,1]$). By Lemma 9.1(a), (9.3) maps U_x conformally onto D. Thus the inverse function $v \mapsto t(v)$ is also holomorphic, and hence $t - 1$ can be expanded in a convergent series

$$(9.4) \qquad t - 1 = \sum_{s=1}^{\infty} c_s v^s, \quad v \in D,$$

where the coefficients c_s are expressible in terms of the p_s; in particular, $c_1 = 1/p_0$.

By Lemma 9.2, one can choose $k \in [0,1) \cap U$, with U as in Lemma 9.1(b), k independent of R and independent of $x \in [0,1]$, such that the disk $|v| \leq |p(k) - p(1)|$ is contained in D for all $x \in [0,1]$. Then the portion $[1,k]$ of $\mathcal{P}$ may be deformed to make its v-map a straight line $[0, \mathcal{K}]$, for all $x \in [0,1]$, without changing the value of the integral $\tilde{I}(R)$ (by (i) and (ii)). Making the change of variable $v = p(t) - p(1)$ on this deformed portion $[1,k]$ of $\mathcal{P}$ we obtain

$$(9.5) \qquad \int_1^k e^{-Rp(t)} \tilde{Q}(t) dt = e^{-Rp(1)} \int_0^{\mathcal{K}} e^{-Rv} f(v) dv,$$

where

$$(9.6) \qquad \mathcal{K} = p(k) - p(1), \quad f(v) = \tilde{Q}(t) \frac{dt}{dv} = \frac{\tilde{Q}(t)}{p'(t)},$$

and the path of integration on the right hand side of (9.5) is a straight line.

By (iii) and (9.4), for $v \in D$, $f(v)$ has a convergent expansion of the form

$$(9.7) \qquad f(v) = \frac{q_0 \left(\sum_{s=1}^{\infty} c_s v^s\right)^{-1/2} + q_1 \left(\sum_{s=1}^{\infty} c_s v^s\right)^{1/2} + \cdots}{p_0 + 2p_1 \left(\sum_{s=1}^{\infty} c_s v^s\right) + 3p_2 \left(\sum_{s=1}^{\infty} c_s v^s\right)^2},$$

the branch of the square root being the principal one. By the binomial theorem one obtains from (9.7) that $f(v)$ can be expressed as a convergent series

$$f(v) = a_0 v^{-1/2} + a_1 v^{1/2} + a_2 v^{3/2} + \cdots,$$

where $v^{1/2}$ has its principal value and the coefficients a_s can be expressed in terms of the p_s's and q_s's as in [28, Ch.3, §8.1]; in particular,

$$(9.8) \qquad a_0 = \frac{q_0}{p_0^{1/2}},$$

the square root having its principal determination.

Define $f_1(v)$ by the relations $f_1(0) = a_1$ and

$$(9.9) \qquad f(v) = a_0 v^{-1/2} + v^{1/2} f_1(v) \quad (v \neq 0).$$

By Lemma 9.3, we may assume that

$$(9.10) \qquad |f_1(v)| \leq M_2, \quad |v| \leq \mathcal{K}, \ x \in [0, 1],$$

where the constant M_2 is independent of $x \in [0, 1]$. (Indeed, by Lemma 9.2, k can be chosen close enough to 1 so that in addition $|\mathcal{K}| = |p(k) - p(1)| < \rho$, $x \in [0, 1]$, for the ρ of Lemma 9.3.)

Using (9.9), rearrange the integral on the right hand side of (9.5) as

$$\int_0^{\mathcal{K}} e^{-Rv} f(v) dv = \frac{a_0}{R^{1/2}} \int_0^\infty e^{-y} y^{-1/2} dy - \frac{a_0}{R^{1/2}} \int_{\mathcal{K}R}^\infty e^{-y} y^{-1/2} dy$$

$$+ \int_0^{\mathcal{K}} e^{-Rv} v^{1/2} f_1(v) dv$$

$$(9.11) \qquad = a_0 R^{-1/2} \Gamma(1/2) - \epsilon_1(R) + \epsilon_2(R),$$

where

$$(9.12) \qquad \epsilon_1(R) = a_0 R^{-1/2} \Gamma(1/2, \mathcal{K}R)$$

$$(9.13) \qquad \epsilon_2(R) = \int_0^{\mathcal{K}} e^{-Rv} v^{1/2} f_1(v) dv.$$

Note that in (9.11) $y^{1/2}$ has its principal value (since $y = Rv$, and $v^{1/2}$ does so) and $\mathcal{K}R$ is not on the negative half-axis (since $\operatorname{Re} \mathcal{K} > 0$ by (v)), so the incomplete Gamma function in (9.12) also takes its principal value.

The absolute value of $\epsilon_1(R)$ can be bounded as follows. By [28, Ch.4, (2.02)] and [28, Ch.4, (2.04)], for real α the incomplete Gamma function $\Gamma(\alpha, z)$ satisfies

$$(9.14) \qquad \Gamma(\alpha, z) = e^{-z} z^{\alpha-1} \{1 + \epsilon(z)\}$$

$$(9.15) \qquad |\epsilon(z)| \leq \frac{|\alpha - 1|}{|z| \cos(\theta - \beta) - \sigma(\beta)},$$

where $\theta = \operatorname{ph} z$ (i.e., $z = re^{i\theta}$ for some $r \geq 0$), $\beta \in (-\pi, \pi)$ is arbitrary,

$$(9.16) \qquad \sigma(\beta) = \sup_{\operatorname{ph} t = -\beta} \frac{\alpha - 2}{|t|} \ln|1 + t|$$

and z is restricted by $|\theta - \beta| < \pi/2$, $|z| \cos(\theta - \beta) > \sigma(\beta)$.

Apply (9.14) and (9.15) for our case, $\alpha = 1/2$, $z = \mathcal{K}R$. Choose $\beta = 0$. We have $\theta = \operatorname{ph} \mathcal{K}R = \operatorname{ph} \mathcal{K}$, so the first condition on z, $|\theta - \beta| < \pi/2$, is met by property (v). The second condition on z is also met, because by (9.16)

$$\sigma(0) = \sup_{t > 0} -\frac{3}{2t} \ln(1 + t) \leq 0.$$

By (9.14) and (9.15) we obtain

$$\Gamma(1/2, \mathcal{K}R) = e^{-\mathcal{K}R} (\mathcal{K}R)^{-1/2} \{1 + \epsilon\},$$

with

$$|\epsilon| \le \frac{1/2}{R|\mathcal{K}|\cos(\mathrm{ph}\,\mathcal{K})} = \frac{1}{2R(-q\ln k)},$$

for all $R \ge 0$. Since $\mathrm{Re}\,\mathcal{K} = -q\ln k > 0$ is independent of x, by the last two relations it follows that

(9.17) $$|\Gamma(1/2, \mathcal{K}R)| \le M_1 R^{-1}, \quad R \ge r_1, \ x \in [0,1],$$

for some constants M_1 and r_1 independent of x. On the other hand, by (9.8) and (iii)

$$a_0 = \frac{(4-2x)^a}{(4-x)^b}\left(-q - i\sqrt{\frac{x}{4-x}}\right)^{-1/2},$$

so $|a_0|$ can clearly be bounded above uniformly for $x \in [0,1]$. By (9.17) and (9.12) it follows that

(9.18) $$|\epsilon_1(R)| \le M_1' R^{-3/2}, \quad R \ge r_1,$$

for all $x \in [0,1]$, where the constants M_1' and r_1 are independent of x.

Next, we bound the absolute value of $\epsilon_2(R)$. The substitution $v = \mathcal{K}\tau$ in (9.13) implies

(9.19) $$\epsilon_2(R) = \mathcal{K}^{3/2}\int_0^1 e^{-R\mathcal{K}\tau}\tau^{1/2}f_1(\mathcal{K}\tau)d\tau.$$

We have

$$\mathrm{Re}\{-R\mathcal{K}\tau\} = -R\tau\,\mathrm{Re}\{p(k) - p(1)\} = -R\tau\eta_k,$$

where $\eta_k = -q\ln k > 0$ is independent of $x \in [0,1]$. Using this and (9.10) we deduce

(9.20) $$|\epsilon_2(R)| \le M_2|\mathcal{K}|^{3/2}\int_0^\infty e^{-R\eta_k\tau}\tau^{1/2}d\tau = M_2|\mathcal{K}|^{3/2}(R\eta_k)^{-3/2}\Gamma(3/2).$$

Since

$$|\mathcal{K}| = |p(k) - p(1)| \le -q\ln k + |\arccos(1 - xk/2) - \arccos(1 - x/2)|$$

and the two functions inside the absolute value sign are continuous in $x \in [0,1]$, $\mathcal{K}$ can be bounded uniformly and (9.20) implies

(9.21) $$|\epsilon_2(R)| \le M_2' R^{-3/2}, \quad R \ge 0,$$

for all $x \in [0,1]$, where M_2' is independent of x.

Substituting (9.18) and (9.21) in (9.11) and using $|e^{-Rp(1)}| = 1$ for all $R \in \mathbb{R}$ and $x \in [0,1]$, we obtain by (9.5)

(9.22) $$\left|\int_1^k e^{-Rp(t)}\tilde{Q}(t)dt - \sqrt{\pi}a_0 R^{-1/2}e^{-Rp(1)}\right| \le M_3 R^{-3/2}, \quad R \ge r_1,$$

for all $x \in [0,1]$, where M_3 and r_1 are constants independent of x.

We now turn to bounding the absolute value of the tail of $\tilde{I}(R)$ omitted by the integral on the left hand sides of (9.5) and (9.22). We have

$$\mathrm{Re}\{p(t) - p(1)\} = -q\ln t \ge \eta > 0, \quad t \in [k, 0),$$

for all $x \in [0, 1]$, where η is independent of x. Therefore for $R \geq r_0$ one has

$$\operatorname{Re}\{Rp(t) - Rp(1)\} = \{(R - r_0) + r_0\} \operatorname{Re}\{p(t) - p(1)\}$$
$$\geq (R - r_0)\eta + \operatorname{Re}\{r_0 p(t) - r_0 p(1)\}.$$

We obtain

$$\left| \int_k^0 e^{-Rp(t)} \tilde{Q}(t) dt \right| \leq |e^{-Rp(1)}| \int_k^0 e^{-\operatorname{Re}\{Rp(t) - Rp(1)\}} |\tilde{Q}(t)| dt$$

$$\leq e^{(r_0 - R)\eta} \int_k^0 |e^{-r_0 p(t)}| |\tilde{Q}(t)| dt$$

(we also used $|e^{-Rp(1)}| = 1$ and $|e^{r_0 p(1)}| = 1$). Choosing $r_0 = -l/q$, since $|e^{-r_0 p(t)}| = t^{q r_0}$, the last integral is uniformly bounded for $x \in [0, 1]$ (see property (iv)). We deduce

$$(9.23) \qquad \left| \int_k^0 e^{-Rp(t)} \tilde{Q}(t) dt \right| \leq M_4 R^{-3/2}, \quad R \geq r_2$$

for all $x \in [0, 1]$, where the constants M_4 and r_2 do not depend on x.

By (9.22) and (9.23) it follows that

$$\left| \tilde{I}(R) - \sqrt{\pi} a_0 R^{-1/2} e^{-Rp(1)} \right| \leq M R^{-3/2}, \quad R \geq R_0,$$

uniformly for $x \in [0, 1]$. This can be rewritten by (9.1) as

$$(9.24) \qquad \left| I(R) - (-i)\sqrt{\pi} a_0 R^{-1/2} e^{-Rp(1)} \right| \leq M R^{-3/2}, \quad R \geq R_0,$$

for all $x \in [0, 1]$, where the constants M and R_0 do not depend on x.

However, by (9.8) and (iii)

$$a_0 = \frac{(4 - 2x)^a}{(4 - x)^b} \left(-q - i\sqrt{\frac{x}{4 - x}} \right)^{-1/2},$$

where the phase of the quantity in the large parentheses has its principal value $-\pi + \arctan 1/q \sqrt{x/(4 - x)}$. This implies

$$a_0 = \frac{(4 - 2x)^a}{(4 - x)^b} \left(-q - i\sqrt{\frac{x}{4 - x}} \right)^{-1/4} e^{i\left[\frac{\pi}{2} - \frac{1}{2} \arctan \frac{1}{q} \sqrt{\frac{x}{4 - x}} \right]}.$$

This shows that the approximant of $I(R)$ in (9.24) has precisely the expression (7.6), and the proof of Proposition 7.2 is complete.

LEMMA 9.1. *Let $t \mapsto v_x(t) = v(t)$ be the map defined by (9.3), all multivalued maps in (9.3) taking their principal values. Let $x \in [0, 1]$ be arbitrary.*

(a). There exists a disk D centered at 0 in the v-plane, independent of $x \in [0, 1]$, and a neighborhood U_x containing 1 in the t-plane, so that $t \mapsto v_x(t)$ maps U_x univalently onto $D(0, \rho)$, and $v_x'(t) \neq 0$ on U_x.

(b). D and U_x can be chosen in part (a) so that there exists a disk U centered at 1 in the t-plane, independent of x, with $U \subset U_x$, for all $x \in [0, 1]$.

PROOF. We have

$$(9.25) \qquad \frac{d}{dt} v_x(t) = -\frac{q}{t} - i\sqrt{\frac{x}{4t - xt^2}} \neq 0, \quad x \in [0, 1], \ |t - 1| < 1,$$

so the last condition in the statement of part (a) is met. To complete the proof of part (a) it suffices to find $\delta, \rho > 0$ independent of $x \in [0,1]$ so that[7]

$$(9.26) \qquad v_x(t) - v_0 = 0 \quad \text{has a unique solution in the disk } |t-1| < \delta,$$
$$\text{for any } v_0 \in D = D(0,\rho).$$

Suppose $\delta > 0$ is independent of $x \in [0,1]$ and satisfies
(9.27)
$$v_x(t) = 0 \text{ has exactly one root in } |t-1| \leq \delta, \text{ for all } x \in [0,1] \text{ (namely } t = 1).$$

Since the map $(x,t) \mapsto |v_x(t)|$ is continuous and non-zero on the compact set $[0,1] \times \{|t-1| = \delta\}$, there exists $m > 0$ independent of $x \in [0,1]$ so that

$$|v_x(t)| \geq m > 0, \ |t-1| = \delta, \ x \in [0,1].$$

Choose ρ so that $0 < \rho < m$. Then for any $v_0 \in D(0,\rho)$ we have $|v_0| < \rho < m \leq |v_x(t)|$, for $|t-1| = \delta$, and Rouché's theorem (see e.g. [1]) implies that $v_x(t)$ and $v_x(t) - v_0$ have the same number of roots inside $|t-1| = \delta$. By (9.27) we obtain that the δ of (9.27) and our choice of ρ satisfy (9.26).

To finish the proof of part (a) we need to prove the existence of some δ satisfying (9.27).

Let $x_0 \in [0,1]$ be fixed. Since $v_{x_0}(1) = 0$, and by (9.25) $v'_{x_0}(1) \neq 0$, there exists $\delta_{x_0} > 0$ so that $v_{x_0}(t)$ has a unique root in $|t-1| \leq \delta_{x_0}$. Set $b_{x_0} = \min_{|t-1|=\delta_{x_0}} |v_{x_0}(t)| > 0$. Write

$$(9.28) \qquad v_x(t) = v_{x_0}(t) + (v_x(t) - v_{x_0}(t)).$$

We have

$$v_x(t) - v_{x_0}(t) = -i[\arccos(1 - xt/2) - \arccos(1 - x_0 t/2)]$$
$$(9.29) \qquad\qquad + i[\arccos(1 - x/2) - \arccos(1 - x_0/2)].$$

Let $f(u) = \arccos(1 - u/2)$.

Consider first the case $x_0 \neq 0$. Since $f'(u) = (4u - u^2)^{-1/2}$, its absolute value is bounded as long as u is bounded away from 0. By the mean value theorem, it follows from (9.29) that there exists an open interval U_{x_0} containing x_0 so that $|v_x(t) - v_{x_0}(t)| < b_{x_0}$ for $x \in U_{x_0}$ and $|t-1| = \delta_{x_0}$. Rouché's theorem applied to (9.28) shows then that $v_x(t)$ has a unique root in $|t-1| \leq \delta_{x_0}$ (namely, $t = 1$) for all $x \in U_{x_0}$.

For $x_0 = 0$ we have

$$v_x(t) - v_{x_0}(t) = -i[\arccos(1 - xt/2) - \arccos(1 - x/2)],$$

and by the continuity of $f(u)$ at $u = 0$ one obtains again that Rouché's theorem is applicable, and there exists a neighborhood U_0 of $x_0 = 0$ in $[0,1]$ so that $v_x(t)$ has a unique root in $|t-1| \leq \delta_{x_0} = \delta_0$ for all $x \in U_0$, namely the root $t = 1$.

Since $[0,1]$ is a compact set, it is covered by a finite subcollection of $(U_x)_{x \in [0,1]}$, say $U_{x_1}, \ldots, U_{x_n}$. Then $\delta = \min_{i=1}^{n} \delta_{x_i}$ satisfies (9.27).

---

[7]The fact that in (9.26) δ is independent of x is not necessary here, but will be needed in the proof of Lemma 9.3.

To prove part (b), consider the inverse function $v \mapsto t_x(v)$, $v \in D$. Let δ and ρ be as in part (a). Since the original domain of (9.3) is the disk $|t - 1| < 1$, we clearly have $\delta < 1$. By (9.25),

$$\frac{dt_x}{dv}(v_x(t)) = \frac{1}{-\frac{q}{t} - i\sqrt{\frac{x}{4t - xt^2}}}, \quad x \in [0, 1], \ |t - 1| < \delta.$$

The denominator above is bounded for x and t as indicated. Furthermore, by part (a), for any fixed $x \in [0, 1]$, the range of $v_x(t)$, $|t - 1| < \delta$, contains the disk $|v| < \rho$. It follows that there exists $\lambda > 0$ independent of x so that $|\frac{dt_x}{dt}(v)| \geq \lambda$, for all $|v| < \rho$ and $x \in [0, 1]$.

For any $0 < \eta < \rho$, consider the circle C_η in the v-plane centered at 0 and having radius η. Since $t_x(v)$ is a conformal map, it follows that $t_x(C_\eta)$ is a simple closed curve in the t-plane containing $t = 1$ in its interior. By the above lower bound on $|\frac{dt_x}{dt}(v)|$, all points of $t_x(C_\eta)$ are at least at distance $\lambda\eta$ from $t = 1$. Therefore, $t_x(C_\eta)$ must contain the disk $|t - 1| < \lambda\eta$ in its interior. Thus, $U_x = t_x(D) = t_x(\cup_{0 \leq \eta < \rho} C_\eta)$ contains the disk $|t - 1| < \lambda\rho$, for all $x \in [0, 1]$. □

LEMMA 9.2. *Let the function $p(t)$ be given by (7.3). Then for any $\rho > 0$ one can choose $k \in [0, 1)$, k not depending on $x \in [0, 1]$, so that the disk $|v| \leq |p(k) - p(1)|$ is contained in $D(0, \rho)$, for all $x \in [0, 1]$.*

PROOF. We have

$$p(k) - p(1) = -q \ln t - i[\arccos(1 - xk/2) - \arccos(1 - x/2)],$$

and since the first term on the right hand side is independent of x, it suffices to show that there exists some $k \in [0, 1)$ so that

(9.30) $|\arccos(1 - xk/2) - \arccos(1 - x/2)| < \rho/2, \quad x \in [0, 1]$.

For $f(u) = \arccos(1 - u/2)$ one has $f'(u) = (4u - u^2)^{-1/2}$, $u \in (0, 1]$, and thus $f(u)$ is increasing on $[0, 1]$. Therefore the absolute value in (9.30) equals in fact

$$g(x) := \arccos(1 - x/2) - \arccos(1 - xk/2).$$

One readily sees that $g'(x) = (4x - x^2)^{-1/2} - (4x/k - x^2)^{-1/2} > 0$, $k \in (0, 1)$, $x \in [0, 1]$, so (9.30) is equivalent to $g(1) = \arccos(1 - 1/2) - \arccos(1 - k/2) < \rho/2$. Since one can clearly choose k to satisfy the latter condition, the proof of the Lemma is complete. □

LEMMA 9.3. *Let $f_1(v)$ be defined by (9.9), where $f(v)$ is given by (9.6), (9.2) and (7.3), and all involved multiple valued maps have their principal determination. Then there exists $\rho > 0$ and a constant M, both independent of $x \in [0, 1]$, so that*

$$|f_1(v)| \leq M, \quad \text{all } |v| < \rho, \ x \in [0, 1].$$

PROOF. By (9.26) and the proof of Lemma 9.1, there exist $0 < \delta < 1$ and $\rho > 0$, both independent of x, so that for any fixed $x \in [0, 1]$, all z's in the v-plane with $|z| < \rho$ can be written as $z = v(t)$, with $|t - 1| < \delta$. Therefore, to prove the current Lemma it is enough to show that $|f_1(v)| = |f_1(v(t))| \leq M$, for $|t - 1| < \delta$ and $x \in [0, 1]$, for some constant M independent of x.

By (9.9) and (9.8) we have

$$
\begin{aligned}
f_1(v) &= v^{-1/2} \left\{ \frac{\tilde{Q}(t)}{p'(t)} - a_0 v^{-1/2} \right\} \\
&= v^{-1/2} \left\{ \frac{H(t)}{p'(t)} (t-1)^{-1/2} - \frac{H(1)}{p'(1)^{1/2}} v^{-1/2} \right\},
\end{aligned}
$$

(9.31)

where

(9.32)
$$
H(t) = (t-1)^{1/2} \tilde{Q}(t).
$$

Write

(9.33)
$$
v = \frac{p(t) - p(1)}{t-1} (t-1) = p'(\xi)(t-1),
$$

where ξ is on the line segment $[1, t]$. Then (9.31) becomes

$$
\begin{aligned}
f_1(v) &= v^{-1/2}(t-1)^{-1/2} \left\{ \frac{H(t)}{p'(t)} - \frac{H(1)}{p'(1)^{1/2} p'(\xi)^{1/2}} \right\} \\
&= \frac{1}{(t-1)p'(\xi)^{1/2}} \left\{ \left(\frac{H(t)}{p'(t)} - \frac{H(1)}{p'(t)} \right) + \left(\frac{H(1)}{p'(t)} - \frac{H(1)}{p'(t)^{1/2} p'(1)^{1/2}} \right) \right. \\
&\qquad\qquad\qquad \left. + \left(\frac{H(1)}{p'(t)^{1/2} p'(1)^{1/2}} - \frac{H(1)}{p'(\xi)^{1/2} p'(1)^{1/2}} \right) \right\} \\
&= \frac{1}{(t-1)p'(\xi)^{1/2}} \left\{ \frac{1}{p'(t)}[H(t) - H(1)] + \frac{H(1)}{p'(t)^{1/2}}[U(t) - U(1)] \right.
\end{aligned}
$$

(9.34)

$$
\left. + \frac{H(1)}{p'(1)^{1/2}}[U(t) - U(\xi)] \right\},
$$

where

(9.35)
$$
U(t) = \frac{1}{p'(t)^{1/2}}.
$$

We have

$$
\begin{aligned}
H(t) - H(1) &= (t-1)H'(\xi_1) \\
U(t) - U(1) &= (t-1)U'(\xi_2) \\
U(t) - U(\xi) &= (t-\xi)U'(\eta),
\end{aligned}
$$

where ξ_1, ξ_2 and η are on the line segment $[1, t]$ (the latter because ξ is on this segment). Therefore (9.34) implies

$$
\begin{aligned}
|f_1(v)| &\leq \frac{1}{|t-1||p'(\xi)^{1/2}|} \left\{ \frac{1}{|p'(t)|}|H(t) - H(1)| + \frac{|H(1)|}{|p'(t)^{1/2}|}|U(t) - U(1)| \right. \\
&\qquad\qquad\qquad \left. + \frac{|H(1)|}{|p'(1)^{1/2}|}|U(t) - U(\xi)| \right\} \\
&= \frac{1}{|t-1||p'(\xi)^{1/2}|} \left\{ \frac{1}{|p'(t)|}|H'(\xi_1)||t-1| + \frac{|H(1)|}{|p'(t)^{1/2}|}|U'(\xi_2)||t-1| \right. \\
&\qquad\qquad\qquad \left. + \frac{|H(1)|}{|p'(1)^{1/2}|}|U'(\eta)||t-\xi| \right\} \\
&\leq \frac{1}{|p'(\xi)^{1/2}|} \left\{ \frac{1}{|p'(t)|}|H'(\xi_1)| + \frac{|H(1)|}{|p'(t)^{1/2}|}|U'(\xi_2)| + \frac{|H(1)|}{|p'(1)^{1/2}|}|U'(\eta)| \right\}
\end{aligned}
$$

(where at the last step we used that ξ is on the line segment $[1, t]$). Therefore, to finish the proof we need upper bounds that are uniform in $x \in [0, 1]$ for $|H(1)|$ and for $|p'(t)|^{-1}$, $|H'(t)|$, and $|U'(t)|$, for $|t - 1| < \delta$.

By (9.32) and (7.4), we have

$$(9.36) \qquad\qquad H(t) = t^l(4 - 2xt)^a/(4 - xt)^b,$$

where $0 \leq a \in \mathbb{Z}$, $-1/2 \leq b \in \frac{1}{2}\mathbb{Z}$ and $l \in \frac{1}{2}\mathbb{Z}$. Since the right hand side of (9.36) is continuous in (x, t) on the compact set $[0, 1] \times \{|t-1| \leq \delta\}$, $|H(t)|$, and in particular $|H(1)|$ can be majorized uniformly with respect to $x \in [0, 1]$.

Since by (9.35) $U'(t) = -1/2[p'(t)]^{-3/2}p''(t)$, all remaining uniform upper bounds will follow from such bounds for $|H'(t)|$, $1/|p'(t)|$ and $|p''(t)|$.

By (9.36) we have
(9.37)
$$H'(t) = lt^{l-1}\frac{(4 - 2xt)^a}{(4 - xt)^b} + t^l \frac{-2ax(4 - 2xt)^{a-1}(4 - xt)^b + bx(4 - 2xt)^a(4 - xt)^{b-1}}{(4 - xt)^{2b}}.$$

It follows from (7.3) that

$$(9.38) \qquad\qquad p'(t) = -\frac{q}{t} - i\sqrt{\frac{x}{4t - xt^2}}$$

$$(9.39) \qquad\qquad p''(t) = \frac{q}{t^2} + i\frac{\sqrt{x}(2 - xt)}{(4t - xt^2)^{3/2}}.$$

Since the right hand sides in (9.37) and (9.39) are continuous in (x, t) on the compact set $[0, 1] \times \{|t-1| \leq \delta\}$, it follows that $|H'(t)|$ and $|p''(t)|$ can be majorized as desired. Since $|t - 1| \leq \delta$, and by (9.38) $|p'(t)| \geq q/|t|$, one has $1/|p'(t)| \leq (1+\delta)/q$, and the proof is complete. $\qquad\square$

The asymptotics of a
multidimensional Laplace integral

By Proposition 8.1, the asymptotics of the moments $M_{\alpha_1,\beta_1,\ldots,\alpha_m,\beta_m;\gamma_1,\delta_1,\ldots,\gamma_n,\delta_n}$ is given by the multiple integral in (8.3). In determining the asymptotics of the latter we will use the following result.

PROPOSITION 10.1. *Let* $h, a : D \to \mathbb{C}$, $(0,1]^n \subset D \subset \mathbb{R}^n$, *be two functions. Assume that* $\int_{[0,1]^n} |h| < \infty$ *and that there exists a neighborhood* V *of* $(1,\ldots,1)$ *so that* $h \in C^1(V)$ *and* $a \in C^2(V)$. *Then for fixed* $q_1,\ldots,q_n > 0$ *we have*

$$I(R) = \int_0^1 \cdots \int_0^1 x_1^{Rq_1} \cdots x_n^{Rq_n} h(x_1,\ldots,x_n) e^{iRa(x_1,\ldots,x_n)} dx_1 \cdots dx_n$$

(10.1)

$$= \frac{1}{R^n} \frac{h(1,\ldots,1) e^{iRa(1,\ldots,1)}}{[q_1 + ia_{x_1}(1,\ldots,1)] \cdots [q_n + ia_{x_n}(1,\ldots,1)]} + O(R^{-n-1}),$$

where a_{x_i} *is the partial derivative of* a *with respect to the variable* x_i.

We deduce the above result from the following.

LEMMA 10.2. *Let* $k \in (0,1)$ *and let* $V \subset \mathbb{R}^n$ *be an open set containing* $[k,1]^n$. *Let* $h, a : V \to \mathbb{C}$ *be functions so that* $h \in C^1(V)$ *and* $a \in C^2(V)$. *Then for fixed* $q_1,\ldots,q_n > 0$ *we have*

$$I_k(R) = \int_k^1 \cdots \int_k^1 x_1^{Rq_1} \cdots x_n^{Rq_n} h(x_1,\ldots,x_n) e^{iRa(x_1,\ldots,x_n)} dx_1 \cdots dx_n$$

$$= \frac{1}{R} \int_k^1 \cdots \int_k^1 \frac{x_2^{Rq_2} \cdots x_n^{Rq_n} h(1, x_2 \ldots, x_n)}{q_1 + ia_{x_1}(1, x_2, \ldots, x_n)} e^{iRa(1,x_2,\ldots,x_n)} dx_2 \cdots dx_n$$

(10.2)

$$+ O(R^{-n-1}).$$

PROOF. Write $I_k(R)$ as

(10.3) $$I_k(R) = \int_k^1 \cdots \int_k^1 h(x_1,\ldots,x_n) e^{Rb(x_1,\ldots,x_n)} dx_1 \cdots dx_n,$$

where

(10.4) $$b(x_1,\ldots,x_n) = q_1 \ln x_1 + \cdots + q_n \ln x_n + ia(x_1,\ldots,x_n).$$

Apply integration by parts[8] with respect to the variable x_1 in (10.3). We have

$$(10.5) \qquad \frac{\partial}{\partial x_1} e^{Rb(x_1,\ldots,x_n)} = R\left(\frac{\partial}{\partial x_1} b(x_1,\ldots,x_n)\right) e^{Rb(x_1,\ldots,x_n)}.$$

By (10.4),

$$(10.6) \qquad \frac{\partial}{\partial x_1} b(x_1,\ldots,x_n) = \frac{q_1}{x_1} + i\frac{\partial}{\partial x_1} a(x_1,\ldots,x_n),$$

so in particular $b_{x_1}(x_1,\ldots,x_n) \neq 0$ for all $(x_1,\ldots,x_n) \in [k,1]^n$. Therefore (10.5) can be rewritten (omitting the arguments for brevity) as

$$(10.7) \qquad e^{Rb} = \frac{1}{Rb_{x_1}} \frac{\partial}{\partial x_1} e^{Rb}.$$

Writing the second factor of the integrand in (10.3) as in (10.7) and applying integration by parts with respect to the variable x_1 we obtain

$$\begin{aligned}
I_k(R) &= \int_k^1 \cdots \int_k^1 \left\{ \int_k^1 h \frac{1}{Rb_{x_1}} \frac{\partial e^{Rb}}{\partial x_1} dx_1 \right\} dx_2 \cdots dx_n \\
&= \int_k^1 \cdots \int_k^1 \left\{ \left. \frac{h}{Rb_{x_1}} e^{Rb} \right|_k^1 - \int_k^1 e^{Rb} \frac{\partial}{\partial x_1}\left(\frac{h}{Rb_{x_1}}\right) dx_1 \right\} dx_2 \cdots dx_n \\
&= \frac{1}{R} \int_k^1 \cdots \int_k^1 \frac{h(1,x_2,\ldots,x_n)}{b_{x_1}(1,x_2,\ldots,x_n)} e^{Rb(1,x_2,\ldots,x_n)} dx_2 \cdots dx_n \\
&\quad - \frac{1}{R} \int_k^1 \cdots \int_k^1 k^{Rq_1} x_2^{Rq_2} \cdots x_n^{Rq_n} \frac{h(k,x_2,\ldots,x_n)}{b_{x_1}(k,x_2,\ldots,x_n)} e^{iRa(k,x_2,\ldots,x_n)} dx_2 \cdots dx_n
\end{aligned}$$

$$(10.8)$$

$$- \frac{1}{R} \int_k^1 \cdots \int_k^1 \frac{h_{x_1} b_{x_1} - h b_{x_1 x_1}}{(b_{x_1})^2} e^{Rb} dx_1 \cdots dx_n$$

(where at the last equality we used (10.4) for the second term on the right hand side).

By (10.6),

$$(10.9) \qquad |b_{x_1}(x_1,\ldots,x_n)| \geq \frac{q_1}{x_1} \geq q_1, \quad (x_1,\ldots,x_n) \in [k,1]^n.$$

Therefore, the absolute value of the second term on the right hand side of (10.8) is majorized by

$$\frac{1}{Rq_1} k^{Rq_1} \int_k^1 \cdots \int_k^1 |h(k,x_2,\ldots,x_n)| dx_2 \cdots dx_n,$$

which in turn, by the presence of the exponential and since h is continuous on $[k,1]^n$, is less than $M_1 R^{-n-1}$ for $R \geq r_1$, for some suitable constants M_1 and r_1.

---

[8]The idea of using integration by parts was suggested to the author by András Vasy.

On the other hand, the absolute value of the third term on the right hand side of (10.8) can be majorized, according to (10.9), by

$$\frac{1}{R(q_1)^2} \int_k^1 \cdots \int_k^1 |h_{x_1} b_{x_1} - h b_{x_1 x_1} e^{Rb}| dx_1 \cdots dx_n$$

$$= \frac{1}{R(q_1)^2} \int_k^1 \cdots \int_k^1 x_1^{Rq_1} \cdots x_n^{Rq_n} |h_{x_1} b_{x_1} - h b_{x_1 x_1}| dx_1 \cdots dx_n.$$

Since $h_{x_1} b_{x_1} - h b_{x_1 x_1}$ is continuous on $[k,1]^n$ and

$$\int_{[k,1]^n} x_1^{Rq_1} \cdots x_n^{Rq_n} \le \int_{[0,1]^n} x_1^{Rq_1} \cdots x_n^{Rq_n} = \prod_{i=1}^n (Rq_i + 1)^{-1},$$

there exist constants M_2 and r_2 so that the right hand side of the above equation is majorized by $M_2 R^{-n-1}$, for all $R \ge r_2$.

By (10.8), (10.6) and the last two paragraphs we obtain (10.3). □

PROOF OF PROPOSITION 10.1. Choose $k \in (0,1)$ so that $[k,1]^n \subset V$. We note first that in order to prove the Proposition it suffices to show that $I_k(R)$ has the asymptotics given by the right hand side of (10.1), where $I_k(R)$ is defined by (10.2). Indeed, $I(R) = I_k(R) + J(R)$, where

$$J(R) = \int_{[0,1]^n \setminus [k,1]^n} x_1^{Rq_1} \cdots x_n^{Rq_n} h(x_1, \ldots, x_n) e^{iRa(x_1, \ldots, x_n)} dx_1 \cdots dx_n.$$

Since throughout the integration range above one has $x_i < k$ for at least one index i, denoting $q_0 = \min(q_1, \ldots, q_n) > 0$ we obtain

$$|J(R)| \le \int_{[0,1]^n \setminus [k,1]^n} k^{Rq_0} |h(x_1, \ldots, x_n)| dx_1 \cdots dx_n$$

$$\le k^{Rq_0} \int_{[0,1]^n} |h(x_1, \ldots, x_n)| dx_1 \cdots dx_n.$$

By hypothesis the integral above is finite, so there exist constants M_0 and r_0 so that $|J(R)| \le M_0 R_0^{-n-1}$ for all $R \ge r_0$. Therefore addition of $J(R)$ to $I_k(R)$ does not change its asymptotics.

By Lemma 10.2, the asymptotics of $I_k(R)$ is the same as $1/R$ times the asymptotics of the integral on the right hand side of (10.2). In turn, the latter integral meets the hypotheses of Lemma 10.2. Applying Lemma 10.2 to it we obtain from (10.2) that

$$I_k(R) = \frac{1}{R^2} \int_k^1 \cdots \int_k^1 \frac{x_3^{Rq_3} \cdots x_n^{Rq_n} h(1, 1, x_3 \ldots, x_n)}{[q_1 + i a_{x_1}(1, 1, x_3, \ldots, x_n)][q_2 + i a_{x_2}(1, 1, x_3, \ldots, x_n)]}$$
$$\times e^{iRa(1, 1, x_3, \ldots, x_n)} dx_3 \cdots dx_n + O(R^{-n-1}).$$

Lemma 10.2 can be applied again to the integral on the right hand side above, yielding an $(n-3)$-fold integral that meets its hypotheses. All the successive applications in this manner of Lemma 10.2 lead to multiple integrals to which it is again applicable. After n applications one obtains (10.1). □

By (8.1) and (8.2), the multiple integral (8.3) leads to integrals having the form of $I(R)$, but with the exponential function replaced by a cosine. As indicated below, this type of integrals can easily be handled by Proposition 10.1.

COROLLARY 10.3. *Let $h, a, c : D \to \mathbb{R}$, $(0,1]^n \subset D \subset \mathbb{R}^n$, be three functions. Assume that $\int_{[0,1]^n} |h| < \infty$ and that there exists a neighborhood V of $(1, \ldots, 1)$ so that $h, c \in C^1(V)$ and $a \in C^2(V)$. Then for fixed $q_1, \ldots, q_n > 0$ we have*

$$K(R) = \int_0^1 \cdots \int_0^1 x_1^{Rq_1} \cdots x_n^{Rq_n} h(x_1, \ldots, x_n)$$
$$\times \cos[Ra(x_1, \ldots, x_n) + c(x_1, \ldots, x_n)] dx_1 \cdots dx_n$$

$$= \frac{1}{R^n} \frac{h(1, \ldots, 1)}{\sqrt{(q_1)^2 + (a_{x_1}(1, \ldots, 1))^2} \cdots \sqrt{(q_n)^2 + (a_{x_n}(1, \ldots, 1))^2}}$$
$$\times \cos\left[Ra(1, \ldots, 1) + c(1, \ldots, 1) \right.$$
$$\left. - \arctan \frac{a_{x_1}(1, \ldots, 1)}{q_1} - \cdots - \arctan \frac{a_{x_n}(1, \ldots, 1)}{q_n} \right]$$

(10.10)
$$+ O(R^{-n-1}),$$

where a_{x_i} is the partial derivative of a with respect to the variable x_i.

PROOF. Since h, a and c are real-valued, $\cos(Ra + c) = \operatorname{Re} e^{iRa+ic}$, and thus

$$K(R) =$$
$$\operatorname{Re} \int_0^1 \cdots \int_0^1 x_1^{Rq_1} \cdots x_n^{Rq_n} h(x_1, \ldots, x_n) e^{ic(x_1, \ldots, x_n)} e^{iRa(x_1, \ldots, x_n)} dx_1 \cdots dx_n.$$

Therefore, by Proposition 10.1

$$(10.11) \quad K(R) = \frac{1}{R^n} \operatorname{Re} \frac{h(1, \ldots, 1) e^{i[Ra(1, \ldots, 1) + c(1, \ldots, 1)]}}{[q_1 + ia_{x_1}(1, \ldots, 1)] \cdots [q_n + ia_{x_n}(1, \ldots, 1)]} + O(R^{-n-1}).$$

Let $a_j = a_{x_j}(1, \ldots, 1)$ and write $q_j + ia_j = \sqrt{q_j^2 + a_j^2} (\cos\theta_j + i\sin\theta_j)$, with $\theta_j = \arctan \frac{a_j}{q_j}$, for $j = 1, \ldots, n$. The contribution of the exponential and the denominator to the fraction in (10.11) becomes

$$\frac{e^{i[Ra(1, \ldots, 1) + c(1, \ldots, 1) - \theta_1 - \cdots - \theta_n]}}{\sqrt{q_1^2 + a_1^2} \cdots \sqrt{q_n^2 + a_n^2}}.$$

Substituting this in (10.11) gives (10.10). □

The asymptotics of ω_b. Proof of Theorem 2.2

In this section we use the results of Section 10 to deduce the asymptotics of ω_b from its expression (6.5). To this end, we need to determine first the asymptotics of the moments $M_{\alpha_1,\beta_1,\ldots,\alpha_m,\beta_m;\gamma_1,\delta_1,\ldots,\gamma_n,\delta_n}$ defined by (6.6).

Recall that the arguments of ω_b—and therefore those of the moments M— approach infinity as specified by (2.3). In particular[9], $v_k = q_k R_k + c_k$, $k = 1, \ldots, m$ and $v'_j = q'_j R'_j + c'_j$, $j = 1, \ldots, n$.

Substituting (8.1) and (8.2) into (8.3) we obtain that

$$
M_{\alpha_1,\beta_1,\ldots,\alpha_m,\beta_m;\gamma_1,\delta_1,\ldots,\gamma_n,\delta_n} = \frac{2^{2m}}{\pi^{m+n}} \frac{1}{\prod_{k=1}^m R_k^3 \prod_{j=1}^n R'_j{}^3}
$$

$$
\times \int_0^1 \cdots \int_0^1 \prod_{k=1}^m \frac{1}{\sqrt[4]{q_k^2 + \frac{\prod_{j=1}^n x_{kj} y_{kj}}{4 - \prod_{j=1}^n x_{kj} y_{kj}}} \sqrt[4]{q_k^2 + \frac{\prod_{j=1}^n z_{kj} w_{kj}}{4 - \prod_{j=1}^n z_{kj} w_{kj}}}}
$$

$$
\times \prod_{j=1}^n \frac{\sqrt[4]{q'_j{}^2 + \frac{\prod_{k=1}^m x_{kj} z_{kj}}{4 - \prod_{k=1}^m x_{kj} z_{kj}}} \sqrt[4]{q'_j{}^2 + \frac{\prod_{k=1}^m y_{kj} w_{kj}}{4 - \prod_{k=1}^m y_{kj} w_{kj}}}}{\sqrt{\frac{\prod_{k=1}^m x_{kj} z_{kj}}{4 - \prod_{k=1}^m x_{kj} z_{kj}}} \sqrt{\frac{\prod_{k=1}^m y_{kj} w_{kj}}{4 - \prod_{k=1}^m y_{kj} w_{kj}}}}
$$

$$
\times \prod_{k=1}^m \left(R_k \sqrt{\frac{\prod_{j=1}^n x_{kj} y_{kj}}{4 - \prod_{j=1}^n x_{kj} y_{kj}}} \right)^{\alpha_k} \left(R_k \sqrt{\frac{\prod_{j=1}^n z_{kj} w_{kj}}{4 - \prod_{j=1}^n z_{kj} w_{kj}}} \right)^{\beta_k}
$$

$$
\times \prod_{j=1}^n \left(R'_j \sqrt{\frac{\prod_{k=1}^m x_{kj} z_{kj}}{4 - \prod_{k=1}^m x_{kj} z_{kj}}} \right)^{\gamma_j} \left(R'_j \sqrt{\frac{\prod_{k=1}^m y_{kj} w_{kj}}{4 - \prod_{k=1}^m y_{kj} w_{kj}}} \right)^{\delta_j}
$$

---

[9]In this section doubly indexed variables appear alongside the complex number i. For this reason, we use here k and j as opposed to the more familiar i and j as indices of these doubly indexed variables.

$$\times \prod_{k=1}^{m} \left\{ \cos \left[R_k \arccos \left(1 - \frac{\prod_{j=1}^{n} x_{kj} y_{kj}}{2} \right) \right. \right.$$

$$\left. - \frac{1}{2} \arctan \frac{1}{q_k} \sqrt{\frac{\prod_{j=1}^{n} x_{kj} y_{kj}}{4 - \prod_{j=1}^{n} x_{kj} y_{kj}}} + \frac{\alpha_k \pi}{2} \right]$$

$$\times \cos \left[R_k \arccos \left(1 - \frac{\prod_{j=1}^{n} z_{kj} w_{kj}}{2} \right) \right.$$

$$\left. \left. - \frac{1}{2} \arctan \frac{1}{q_k} \sqrt{\frac{\prod_{j=1}^{n} z_{kj} w_{kj}}{4 - \prod_{j=1}^{n} z_{kj} w_{kj}}} + \frac{\beta_k \pi}{2} \right] \right\}$$

$$\times \prod_{j=1}^{n} \left\{ \cos \left[R'_j \arccos \left(1 - \frac{\prod_{k=1}^{m} x_{kj} z_{kj}}{2} \right) \right. \right.$$

$$\left. + \frac{1}{2} \arctan \frac{1}{q'_j} \sqrt{\frac{\prod_{k=1}^{m} x_{kj} z_{kj}}{4 - \prod_{k=1}^{m} x_{kj} z_{kj}}} + \frac{(\gamma_j - 1)\pi}{2} \right]$$

$$\times \cos \left[R'_j \arccos \left(1 - \frac{\prod_{k=1}^{m} y_{kj} w_{kj}}{2} \right) \right.$$

$$\left. \left. + \frac{1}{2} \arctan \frac{1}{q'_j} \sqrt{\frac{\prod_{k=1}^{m} y_{kj} w_{kj}}{4 - \prod_{k=1}^{m} y_{kj} w_{kj}}} + \frac{(\delta_j - 1)\pi}{2} \right] \right\}$$

$$\times \prod_{k=1}^{m} \prod_{j=1}^{n} (x_{kj} y_{kj} z_{kj} w_{kj})^{q_k R_k + q'_j R'_j + c_k + c'_j + 1} dx_{11} \cdots dw_{mn}$$

$$+ O \left(R^{-4mn + \sum_{i=1}^{m} (\alpha_i + \beta_i) + \sum_{i=1}^{n} (\gamma_i + \delta_i) - 3m - 3n - 1} \right).$$

(11.1)

One readily proves by induction on s that

$$\cos \theta_1 \cdots \cos \theta_s = \frac{1}{2^s} \sum_{\epsilon_1, \dots, \epsilon_s = \pm 1} \cos(\epsilon_1 \theta_1 + \cdots + \epsilon_s \theta_s).$$

Since by (2.3) we have $R_k = A_k R$, $k = 1, \dots, m$ and $R'_j = B_j R$, $j = 1, \dots, n$, the product of the $2m + 2n$ cosines in the integrand of (11.1) becomes by the above formula

$$
\frac{1}{2^{2m+2n}} \sum_{\epsilon_1,\ldots,\epsilon_{2m+2n}=\pm 1} \cos\left\{ R\left[\sum_{k=1}^{m} \left(\epsilon_{2k-1} A_k \arccos\left(1 - \frac{\prod_{j=1}^{n} x_{kj} y_{kj}}{2}\right) \right.\right.\right.
$$

$$
\left.+ \epsilon_{2k} A_k \arccos\left(1 - \frac{\prod_{j=1}^{n} z_{kj} w_{kj}}{2}\right)\right)
$$

$$
+ \sum_{j=1}^{n} \left(\epsilon_{2m+2j-1} B_j \arccos\left(1 - \frac{\prod_{k=1}^{m} x_{kj} z_{kj}}{2}\right)\right.
$$

$$
\left.\left.+ \epsilon_{2m+2j} B_j \arccos\left(1 - \frac{\prod_{k=1}^{m} y_{kj} w_{kj}}{2}\right)\right)\right]
$$

$$
- \sum_{k=1}^{m} \left(\frac{\epsilon_{2k-1}}{2} \arctan \frac{1}{q_k}\sqrt{\frac{\prod_{j=1}^{n} x_{kj} y_{kj}}{4 - \prod_{j=1}^{n} x_{kj} y_{kj}}} + \frac{\epsilon_{2k}}{2} \arctan \frac{1}{q_k}\sqrt{\frac{\prod_{j=1}^{n} z_{kj} w_{kj}}{4 - \prod_{j=1}^{n} z_{kj} w_{kj}}}\right)
$$

$$
+ \sum_{j=1}^{n} \left(\frac{\epsilon_{2m+2j-1}}{2} \arctan \frac{1}{q_j'}\sqrt{\frac{\prod_{k=1}^{m} x_{kj} z_{kj}}{4 - \prod_{k=1}^{m} x_{kj} z_{kj}}}\right.
$$

$$
\left.+ \frac{\epsilon_{2m+2j}}{2} \arctan \frac{1}{q_j'}\sqrt{\frac{\prod_{k=1}^{m} y_{kj} w_{kj}}{4 - \prod_{k=1}^{m} y_{kj} w_{kj}}}\right)
$$

$$
+ \sum_{k=1}^{m} (\epsilon_{2k-1}\alpha_k + \epsilon_{2k}\beta_k)\frac{\pi}{2} + \sum_{j=1}^{n} (\epsilon_{2m+2j-1}\gamma_j + \epsilon_{2m+2j}\delta_j)\frac{\pi}{2}
$$

$$
- \sum_{j=1}^{n} (\epsilon_{2m+2j-1} + \epsilon_{2m+2j})\frac{\pi}{2}\right\}.
$$

(11.2)

The $4mn$-fold integral in (11.1), with the product of cosines in the integrand written as in (11.2), becomes the sum of 2^{2m+2n} integrals of the form of the multiple integral in Corollary 10.3. Our current functions h satisfy $\int_{[0,1]^{4mn}} |h| < \infty$. Indeed, since the product of the $2n$ products (8.12) is $\prod_{k=1}^{m}\prod_{j=1}^{n} x_{kj} y_{kj} z_{kj} w_{kj}$, one readily sees that $|h|$ can be majorized by $K\prod_{k=1}^{m}\prod_{j=1}^{n}(x_{kj} y_{kj} z_{kj} w_{kj})^{c_k + c_j' + 1/2}$, where K is a constant independent of $x_{kj}, y_{kj}, z_{kj}, w_{kj}$. Since by (2.3) the exponent in the previous double product is non-negative, it follows that $\int_{[0,1]^{4mn}} |h| < \infty$.

The other conditions in the hypothesis of Corollary 10.3 are clearly satisfied. The a-functions of Corollary 10.3 are now the coefficients $a(x_{11}, \ldots, w_{mn})$ of R in the argument of the cosines in (11.2).

Note that there are exactly two terms of each of our $a(x_{11}, \ldots, w_{mn})$'s containing any given x_{kj}, y_{kj}, z_{kj} or w_{kj}, and that all variables are set to 1 on the right hand side of (10.10). Since $\partial/\partial x \arccos(1 - xu/2) = (4x/u - x^2)^{-1/2}$, we obtain

$$\frac{\partial a}{\partial x_{kj}}(1, \ldots, 1) = \frac{\epsilon_{2k-1}A_k + \epsilon_{2m+2j-1}B_j}{\sqrt{3}}$$

$$\frac{\partial a}{\partial y_{kj}}(1, \ldots, 1) = \frac{\epsilon_{2k-1}A_k + \epsilon_{2m+2j}B_j}{\sqrt{3}}$$

$$\frac{\partial a}{\partial z_{kj}}(1, \ldots, 1) = \frac{\epsilon_{2k}A_k + \epsilon_{2m+2j-1}B_j}{\sqrt{3}}$$

(11.3)
$$\frac{\partial a}{\partial w_{kj}}(1, \ldots, 1) = \frac{\epsilon_{2k}A_k + \epsilon_{2m+2j}B_j}{\sqrt{3}}.$$

By (11.1)–(11.3), Corollary 10.3 implies

$$M_{\alpha_1,\beta_1,\ldots,\alpha_m,\beta_m;\gamma_1,\delta_1,\ldots,\gamma_n,\delta_n} = \frac{E}{R^{4mn}} \prod_{k=1}^{m} \left(\frac{R_k}{\sqrt{3}}\right)^{\alpha_k+\beta_k} \prod_{j=1}^{n} \left(\frac{R'_j}{\sqrt{3}}\right)^{\gamma_j+\delta_j}$$

$$\times \sum_{\epsilon_1,\ldots,\epsilon_{2m+2n}=\pm 1} \frac{1}{D^{\epsilon_1,\ldots,\epsilon_{2m+2n}}} \cos\left\{\frac{R\pi}{3}\left[\sum_{k=1}^{m} A_k(\epsilon_{2k-1} + \epsilon_{2k})\right.\right.$$

$$\left. + \sum_{j=1}^{n} B_j(\epsilon_{2m+2j-1} + \epsilon_{2m+2j})\right]$$

$$- \sum_{k=1}^{m} \frac{\epsilon_{2k-1} + \epsilon_{2k}}{2} \arctan\frac{1}{q_k\sqrt{3}} + \sum_{j=1}^{n} \frac{\epsilon_{2m+2j-1} + \epsilon_{2m+2j}}{2} \arctan\frac{1}{q'_j\sqrt{3}}$$

$$+ \sum_{k=1}^{m} (\epsilon_{2k-1}\alpha_k + \epsilon_{2k}\beta_k)\frac{\pi}{2} + \sum_{j=1}^{n} (\epsilon_{2m+2j-1}\gamma_j + \epsilon_{2m+2j}\delta_j)\frac{\pi}{2}$$

$$- \sum_{j=1}^{n} (\epsilon_{2m+2j-1} + \epsilon_{2m+2j})\frac{\pi}{2}$$

$$- \sum_{k=1}^{m}\sum_{j=1}^{n} \left(\arctan\frac{\epsilon_{2k-1}A_k + \epsilon_{2m+2j-1}B_j}{(q_kA_k + q'_jB_j)\sqrt{3}} + \arctan\frac{\epsilon_{2k-1}A_k + \epsilon_{2m+2j}B_j}{(q_kA_k + q'_jB_j)\sqrt{3}}\right.$$

$$\left.\left. + \arctan\frac{\epsilon_{2k}A_k + \epsilon_{2m+2j-1}B_j}{(q_kA_k + q'_jB_j)\sqrt{3}} + \arctan\frac{\epsilon_{2k}A_k + \epsilon_{2m+2j}B_j}{(q_kA_k + q'_jB_j)\sqrt{3}}\right)\right\}$$

(11.4)
$$+ O(R^{-4mn+\sum_{k=1}^{m}(\alpha_k+\beta_k)+\sum_{j=1}^{n}(\gamma_j+\delta_j)-3m-3n-1}),$$

where

(11.5)
$$E = \frac{3^n}{2^{2n}\pi^{m+n}} \frac{1}{\prod_{k=1}^{m} R_k^3 \prod_{j=1}^{n} R_j'^3} \frac{\prod_{j=1}^{n} \sqrt{q_j'^2 + \frac{1}{3}}}{\prod_{k=1}^{m} \sqrt{q_k^2 + \frac{1}{3}}}$$

and

$$D^{\epsilon_1,\ldots,\epsilon_{2m+2n}} = \prod_{k=1}^{m}\prod_{j=1}^{n}\left\{\sqrt{(q_k A_k + q'_j B_j)^2 + \frac{(\epsilon_{2k-1}A_k + \epsilon_{2m+2j-1}B_j)^2}{3}}\right.$$

$$\times \sqrt{(q_k A_k + q'_j B_j)^2 + \frac{(\epsilon_{2k-1}A_k + \epsilon_{2m+2j}B_j)^2}{3}}$$

$$\times \sqrt{(q_k A_k + q'_j B_j)^2 + \frac{(\epsilon_{2k}A_k + \epsilon_{2m+2j-1}B_j)^2}{3}}$$

(11.6)
$$\left.\times \sqrt{(q_k A_k + q'_j B_j)^2 + \frac{(\epsilon_{2k}A_k + \epsilon_{2m+2j}B_j)^2}{3}}\right\}.$$

Define

(11.7)
$$S^{\epsilon_1,\ldots,\epsilon_{2m+2n}}_{\alpha_1,\beta_1,\ldots,\alpha_m,\beta_m;\gamma_1,\delta_1,\ldots,\gamma_n,\delta_n} := \prod_{k=1}^{m}\left(\frac{R_k}{\sqrt{3}}\right)^{\alpha_k+\beta_k}\prod_{j=1}^{n}\left(\frac{R'_j}{\sqrt{3}}\right)^{\gamma_j+\delta_j}$$

$$\times \text{(summand of the } (2m+2n)\text{-fold sum in (11.4))}.$$

We call $(\epsilon_1,\ldots,\epsilon_{2m+2n})$ *balanced* if $\epsilon_{2j-1} + \epsilon_{2j} = 0$, $j = 1,\ldots,2m+2n$.

LEMMA 11.1. *Fix $\epsilon_j \in \{-1,1\}$, $j = 1,\ldots,2m+2n$. As in Section 6, let $\mathcal{C}$ be the collection of terms obtained by expanding out the left hand side of (6.2), and for any $C \in \mathcal{C}$ let $e(C)$, $\alpha_k(C)$, $\beta_k(C)$, $\gamma_j(C)$ and $\delta_j(C)$ have the same significance as in (6.2).*

Then unless $(\epsilon_1,\ldots,\epsilon_{2m+2n})$ is balanced, one has

(11.8)
$$\sum_{C\in\mathcal{C}} e(C)S^{\epsilon_1,\ldots,\epsilon_{2m+2n}}_{\alpha_1(C),\beta_1(C),\ldots,\alpha_m(C),\beta_m(C);\gamma_1(C),\delta_1(C),\ldots,\gamma_n(C),\delta_n(C)} = 0.$$

PROOF. Suppose there exists $j \in [1,m]$ with $\epsilon_{2j-1} = \epsilon_{2j}$. Without loss of generality we may assume $\epsilon_1 = \epsilon_2$. Then it follows from (11.7) that

(11.9)
$$S^{\epsilon_1,\ldots,\epsilon_{2m+2n}}_{\alpha_1,\beta_1,\alpha_2,\beta_2,\ldots,\alpha_m,\beta_m;\gamma_1,\delta_1,\ldots,\gamma_n,\delta_n} = S^{\epsilon_1,\ldots,\epsilon_{2m+2n}}_{\alpha_1-1,\beta_1+1,\alpha_2,\beta_2,\ldots,\alpha_m,\beta_m;\gamma_1,\delta_1,\ldots,\gamma_n,\delta_n}, \quad \alpha_1 \geq 1.$$

Partition the terms in the expansion $\mathcal{C}$ of the left hand side of (6.2) into four classes $\mathcal{C}_1$, $\mathcal{C}_2$, $\mathcal{C}_3$, $\mathcal{C}_4$, according to which of the terms of the factor $(a_1 - b_1)^2 = (a_1^2 - a_1 b_1 - b_1 a_1 + b_1^2)$ (in order, from left to right) is chosen when expanding.

By (11.9), the restriction of the sum (11.8) to $\mathcal{C}_1$ is canceled by its restriction to $\mathcal{C}_2$, and the restriction to $\mathcal{C}_3$ is canceled by the restriction to $\mathcal{C}_4$. This proves (11.8).

The case when $\epsilon_{2j-1} = \epsilon_{2j}$, $j \in [m+1,m+n]$, is treated in a perfectly analogous manner. □

Define $M^0_{\alpha_1,\beta_1,\alpha_2,\beta_2,\ldots,\alpha_m,\beta_m;\gamma_1,\delta_1,\ldots,\gamma_n,\delta_n}$ to be given by the expression on the right hand side of (11.4), when the summation range is restricted to balanced $(\epsilon_1,\ldots,\epsilon_{2m+2n})$'s.

Clearly, for $(\epsilon_1,\ldots,\epsilon_{2m+2n})$ balanced we have

$$\{\epsilon_{2k-1}A_k + \epsilon_{2m+2j-1}B_j, \epsilon_{2k-1}A_k + \epsilon_{2m+2j}B_j, \epsilon_{2k}A_k + \epsilon_{2m+2j-1}B_j,$$

$$\epsilon_{2k}A_k + \epsilon_{2m+2j}B_j\}$$

$$= \{A_k + B_j, A_k - B_j, -A_k + B_j, -A_k - B_j\},$$

for all $k = 1, \ldots, m$, $j = 1, \ldots, n$. It follows that for $(\epsilon_1, \ldots, \epsilon_{2m+2n})$ balanced one has

$\quad (i)$ the double sum of arctangents in (11.4) is 0, and
$\quad (ii)$ the double product $D^{\epsilon_1, \ldots, \epsilon_{2m+2n}}$ in (11.6) equals
(11.10)
$$D = \prod_{k=1}^{m} \prod_{j=1}^{n} \left[(q_k A_k + q'_j B_j)^2 + \frac{1}{3}(A_k - B_j)^2 \right] \left[(q_k A_k + q'_j B_j)^2 + \frac{1}{3}(A_k + B_j)^2 \right],$$

an expression independent of $(\epsilon_1, \ldots, \epsilon_{2m+2n})$.

By (i) and (ii) above we obtain from our definition of the M_0's that

$$M^0_{\alpha_1, \beta_1, \alpha_2, \beta_2, \ldots, \alpha_m, \beta_m; \gamma_1, \delta_1, \ldots, \gamma_n, \delta_n} = \frac{E}{DR^{4mn}} \prod_{k=1}^{m} \left(\frac{R_k}{\sqrt{3}} \right)^{\alpha_k + \beta_k} \prod_{j=1}^{n} \left(\frac{R'_j}{\sqrt{3}} \right)^{\gamma_j + \delta_j}$$

(11.11)

$$\times \sum_{\epsilon_{2l-1} = \pm 1, \ l=1, \ldots, m+n} \cos \left(\sum_{k=1}^{m} \frac{\epsilon_{2k-1}(\alpha_k - \beta_k)}{2} \pi + \sum_{j=1}^{n} \frac{\epsilon_{2m+2j-1}(\gamma_j - \delta_j)}{2} \pi \right).$$

By (6.2), when our parameters depend on R as in (2.3), the order in R of $e(C)$ plus $\sum_{k=1}^{m}(\alpha_k(C) + \beta_k(C)) + \sum_{j=1}^{n}(\gamma_j(C) + \delta_j(C))$ is always less or equal than the number of factors of the product $\mathcal{E}$ of (6.2), namely $2m + 2n + 4\binom{m}{2} + 4\binom{n}{2} = 2m^2 + 2n^2$. Thus, the omitted part of the approximation of

$$e(C)M_{\alpha_1(C), \beta_1(C), \ldots, \alpha_m(C), \beta_m(C); \gamma_1(C), \delta_1(C), \ldots, \gamma_n(C), \delta_n(C)}$$

resulting from (11.4) has at most order $2m^2 + 2n^2 - 4mn - 3m - 3n - 1$ in R. Therefore, by Lemma 11.1 and our definition of the M_0's, (6.5) implies

$$\omega_b \left(\begin{matrix} R_1 & \cdots & R_m \\ v_1 & & v_m \end{matrix} ; \begin{matrix} R'_1 & \cdots & R'_n \\ v'_1 & & v'_n \end{matrix} \right) =$$

$$X \left| \sum_{C \in \mathcal{C}} e(C)M^0_{\alpha_1(C), \beta_1(C), \ldots, \alpha_m(C), \beta_m(C); \gamma_1(C), \delta_1(C), \ldots, \gamma_n(C), \delta_n(C)} \right|$$

(11.12)

$$+ O(R^{2m^2 + 2n^2 - 4mn - 2m - 1}),$$

where

(11.13)
$$X = \chi_{2m,2n} \prod_{k=1}^{m} R_k \prod_{j=1}^{n} R'_j (R'_j - 1/2)(R'_j + 1/2),$$

with χ given by (4.3).

LEMMA 11.2. *We have*

$M^0_{\alpha_1,\beta_1,\ldots,\alpha_m,\beta_m;\gamma_1,\delta_1,\ldots,\gamma_n,\delta_n}$

(11.14)

$$
= \begin{cases}
Y \displaystyle\prod_{k=1}^{m} \left(\frac{iR_k}{\sqrt{3}}\right)^{\alpha_k} \prod_{k=1}^{m} \left(\frac{-iR_k}{\sqrt{3}}\right)^{\beta_k} \prod_{j=1}^{n} \left(\frac{iR'_j}{\sqrt{3}}\right)^{\gamma_j} \prod_{j=1}^{n} \left(\frac{-iR'_j}{\sqrt{3}}\right)^{\delta_j}, \\
\qquad\qquad \text{if } \alpha_1 = \beta_1 (\mathrm{mod}\,2), \ldots, \gamma_n = \delta_n (\mathrm{mod}\,2) \\
0, \qquad\qquad \text{otherwise,}
\end{cases}
$$

where

$$
Y = \frac{2^{m-n}3^n}{\pi^{m+n}R^{4mn}} \frac{1}{\prod_{k=1}^{m} R_k^3 \prod_{j=1}^{n}(R'_j)^3} \frac{\prod_{j=1}^{n}\sqrt{{q'_j}^2 + \frac{1}{3}}}{\prod_{k=1}^{m}\sqrt{q_k^2 + \frac{1}{3}}}
$$

(11.15)

$$
\times \frac{1}{\prod_{k=1}^{m}\prod_{j=1}^{n}[(q_k A_k + q'_j B_j)^2 + \frac{1}{3}(A_k - B_j)^2][(q_k A_k + q'_j B_j)^2 + \frac{1}{3}(A_k + B_j)^2]}.
$$

PROOF. If $\alpha_1 = \beta_1 (\mathrm{mod}\,2), \ldots, \gamma_n = \delta_n (\mathrm{mod}\,2)$, all the constants multiplying π in the argument of the cosine in (11.11) are integers. Since $\cos(-n\pi + x) = \cos(n\pi + x)$ for all $n \in \mathbb{Z}$ and all x, it follows that all 2^{m+n} terms in the sum (11.11) are equal to the term corresponding to $(\epsilon_1, \epsilon_3 \ldots, \epsilon_{2m+2n-1}) = (1, \ldots, 1)$, which is

$$
\cos\left(\sum_{k=1}^{m}\frac{\alpha_k - \beta_k}{2}\pi + \sum_{j=1}^{n}\frac{\gamma_j - \delta_j}{2}\pi\right) = (-1)^{\sum_{k=1}^{m}\frac{\alpha_k-\beta_k}{2} + \sum_{j=1}^{n}\frac{\gamma_j-\delta_j}{2}}
$$

$$
= \prod_{k=1}^{m} i^{\alpha_k}(-i)^{\beta_k} \prod_{j=1}^{n} i^{\gamma_j}(-i)^{\delta_j}.
$$

Using this and substituting in (11.11) the expressions for E and D given by (11.5) and (11.10) we obtain (11.14).

Assume next that there is an index k so that α_k and β_k have opposite parity. Since $\cos(-n\pi + x) = -\cos(n\pi + x)$ for all $n \in \frac{1}{2} + \mathbb{Z}$ and all x, it follows that the $\epsilon_{2k-1} = 1$ part of the multiple sum in (11.11) cancels the $\epsilon_{2k-1} = -1$ part, so the multiple sum is 0. The same argument applies if there is an index j so that γ_j and δ_j have opposite parity. □

By (11.12) and Lemma 11.2 we obtain

$$
\omega_b \begin{pmatrix} R_1 & \ldots & R_m & R'_1 & \ldots & R'_n \\ v_1 & & v_m & v'_1 & & v'_n \end{pmatrix} = XY
$$

$$
\times \left| \sum_{C \in \mathcal{C}} e(C) \left\langle \prod_{k=1}^{m}\left(\frac{iR_k}{\sqrt{3}}\right)^{\alpha_k(C)} \prod_{k=1}^{m}\left(\frac{-iR_k}{\sqrt{3}}\right)^{\beta_k(C)} \prod_{j=1}^{n}\left(\frac{iR'_j}{\sqrt{3}}\right)^{\gamma_j(C)} \prod_{j=1}^{n}\left(\frac{-iR'_j}{\sqrt{3}}\right)^{\delta_j(C)} \right\rangle \right|
$$

(11.16)

$$
+ O(R^{2m^2+2n^2-4mn-2m-1}),
$$

where for any monomial $\mu = \prod_{k=1}^{m} x_k^{\alpha_k} y_k^{\beta_k} \prod_{j=1}^{n} z_j^{\gamma_j} w_j^{\delta_j}$, the angular brackets denote $\langle \mu \rangle = \mu$ in case $\alpha_k = \beta_k \pmod 2$, $\gamma_j = \delta_j \pmod 2$, for all k and j, and $\langle \mu \rangle = 0$ otherwise.

Note that if in (11.16) we had $\langle \mu \rangle = \mu$ for *all* monomials μ, the sum in (11.16) would evaluate to the specialization of the expression $\mathcal{E}$ of (6.2) when

$$(11.17) \qquad a_k = \frac{iR_k}{\sqrt{3}}, \ b_k = -\frac{iR_k}{\sqrt{3}}, \ c_j = \frac{iR'_j}{\sqrt{3}}, \ d_j = -\frac{iR'_j}{\sqrt{3}},$$

$k = 1, \ldots, m$, $j = 1, \ldots, n$.

This observation, together with the fortunate situation presented in the result below, allows us to evaluate the sum (11.16).

LEMMA 11.3. *Under the substitutions* (11.17), *all contributions to the expansion of $\mathcal{E}$ coming from monomials*

$$\mu_C = e(C) a_1^{\alpha_1(C)} b_1^{\beta_1(C)} \cdots a_m^{\alpha_m(C)} b_m^{\beta_m(C)} c_1^{\gamma_1(C)} d_1^{\delta_1(C)} \cdots c_n^{\gamma_n(C)} d_n^{\delta_n(C)},$$

$C \in \mathcal{C}$, *for which not all pairs* $(\alpha_k(C), \beta_k(C))$, $(\gamma_j(C), \delta_j(C))$ *have components of the same parity, cancel out.*

PROOF. We define first the following map $C \mapsto C'$ on the subcollection $\mathcal{C}_0$ consisting of those $C \in \mathcal{C}$ for which not all pairs $(\alpha_k(C), \beta_k(C))$, $(\gamma_j(C), \delta_j(C))$ have components of the same parity.

Totally order the disjoint union of the set of indices in the pairs $(\alpha_k(C), \beta_k(C))$ with the set of indices in the pairs $(\gamma_j(C), \delta_j(C))$. Let $C \in \mathcal{C}_0$, and consider the smallest pair index for which the components have opposite parity. Assume that this smallest index k_0 occurs in a pair of the first type, $(\alpha_{k_0}(C), \beta_{k_0}(C))$. Our arguments apply the same way to the case when the smallest such index occurs in a pair of type $(\gamma_j(C), \delta_j(C))$.

By (6.2), C is obtained by selecting a signed term from each of the $2m + 2n + 4\binom{m}{2} + 4\binom{n}{2}$ factors of $\mathcal{E}$. Define C' as being obtained by making the following selections:

(*i*) From the m pairs of factors $(a_k - b_k)(a_k - b_k)$, $k = 1, \ldots, m$, make the selection as follows:

    (a) if $l \neq k_0$, select in C' the same signed terms as in C;

    (b) if $k = k_0$, select in C' the signed term of each factor of $(a_{k_0} - b_{k_0})(a_{k_0} - b_{k_0})$ that was *not* selected in C.

(*ii*) From the n pairs of factors $(c_j - d_j)(c_j - d_j)$, $j = 1, \ldots, n$, select in C' the same signed terms as in C.

(*iii*) From the $\binom{m}{2}$ groups of four factors

$$((v_j - v_k) + a_j - a_k)((v_j - v_k) + a_j - b_k)((v_j - v_k) + b_j - a_k)((v_j - v_k) + b_j - b_k),$$

$1 \leq k < j \leq m$, make the selection as follows:

    (a) if $k_0 \notin \{k, j\}$, make in C' the same selection as in C;

    (b) if $k = k_0$, and the selected terms in C from the four factors

$$((v_j - v_{k_0}) + a_j - a_{k_0})((v_j - v_{k_0}) + a_j - b_{k_0})((v_j - v_{k_0}) + b_j - a_{k_0})((v_j - v_{k_0}) + b_j - b_{k_0})$$

are the k_1th, k_2th, k_3th and k_4th, respectively, then select in C' the k_2th, k_1th, k_4th and k_3th terms of the above factors, respectively;

(c) if $j = k_0$, and the selected terms in C from the four factors

$$((v_{k_0}-v_k)+a_{k_0}-a_k)((v_{k_0}-v_k)+a_{k_0}-b_k)((v_{k_0}-v_k)+b_{k_0}-a_k)((v_{k_0}-v_k)+b_{k_0}-b_k)$$

are the k_1th, k_2th, k_3th and k_4th, respectively, then select in C' the k_3th, k_4th, k_1th and k_2th terms of the above factors, respectively.

(iv) From the final $\binom{n}{2}$ groups of four factors make the same selections in C' as in C.

Let us now compare the monomials

$$\mu_C = e(C) \prod_{k=1}^m a_k^{\alpha_k(C)} b_k^{\beta_k(C)} \prod_{j=1}^m c_j^{\gamma_j(C)} d_j^{\delta_j(C)}$$

and

$$\mu_{C'} = e(C') \prod_{k=1}^m a_k^{\alpha_k(C')} b_k^{\beta_k(C')} \prod_{j=1}^m c_j^{\gamma_j(C')} d_j^{\delta_j(C')}$$

generated by the selections C and C'.

It is clear from our construction that the portions of C and C' covered by step (iii) produce contributions to μ_C and $\mu_{C'}$ whose only difference is that the roles of a_{k_0} and b_{k_0} are interchanged.

The same is true for step (i). Indeed, for $k = k_0$, to the four possible selections $(a_{k_0})(a_{k_0})$, $(a_{k_0})(-b_{k_0})$, $(-b_{k_0})(a_{k_0})$, $(-b_{k_0})(-b_{k_0})$ in C there correspond the four selections $(-b_{k_0})(-b_{k_0})$, $(-b_{k_0})(a_{k_0})$, $(a_{k_0})(-b_{k_0})$, $(a_{k_0})(a_{k_0})$ in C', respectively. So the four possible contributions to μ_C and $\mu_{C'}$ are $a_{k_0}^2 b_{k_0}^0$, $-a_{k_0} b_{k_0}$, $-a_{k_0}^2 b_{k_0}$, $a_{k_0}^0 b_{k_0}^2$ and $a_{k_0}^0 b_{k_0}^2$, $-a_{k_0} b_{k_0}$, $-a_{k_0}^2 b_{k_0}$, $a_{k_0}^2 b_{k_0}^0$, respectively.

Since steps (ii) and (iv) do not involve the index k_0, and the portions of C and C' covered by these steps coincide, it follows from the previous two paragraphs that

$$\mu_C(a_1,b_1,\ldots,a_{k_0},b_{k_0},\ldots,a_m,b_m,c_1,d_1,\ldots,c_n,d_n)$$
$$= \mu_{C'}(a_1,b_1,\ldots,b_{k_0},a_{k_0},\ldots,a_m,b_m,c_1,d_1,\ldots,c_n,d_n).$$

Since by (11.17) a_{k_0} and b_{k_0} become the negatives of each other, and since $\alpha_{k_0}(C)$ and $\beta_{k_0}(C)$ have opposite parity, it follows that under the specialization (11.17) the monomials μ_C and $\mu_{C'}$ cancel out.

Our map $C \mapsto C'$ clearly maps $\mathcal{C}_0$ to itself, and sends C' back to C. By the previous paragraph, when making the specialization (11.17), all the monomials corresponding to term selections in $\mathcal{C}_0$ cancel out in pairs. This proves the Lemma. $\square$

COROLLARY 11.4. *We have*

$$\omega_b \begin{pmatrix} R_1 & \cdots & R_m \\ v_1 & & v_m \end{pmatrix}; \begin{matrix} R'_1 & \cdots & R'_n \\ v'_1 & & v'_n \end{matrix} = XY$$

$$\times \left| \sum_{C \in \mathcal{C}} e(C) \prod_{k=1}^m \left(\frac{iR_k}{\sqrt{3}}\right)^{\alpha_k(C)} \prod_{k=1}^m \left(\frac{-iR_k}{\sqrt{3}}\right)^{\beta_k(C)} \prod_{j=1}^n \left(\frac{iR'_j}{\sqrt{3}}\right)^{\gamma_j(C)} \prod_{j=1}^n \left(\frac{-iR'_j}{\sqrt{3}}\right)^{\delta_j(C)} \right|$$

(11.18)
$$+ O(R^{2m^2+2n^2-4mn-2m-1}),$$

where X and Y are given by (11.13) and (11.15).

PROOF. This follows directly from (11.16) and Lemma 11.3.   □

Our results allow us now to obtain the asymptotics of ω_b, thus proving Theorem 2.2.

PROOF OF THEOREM 2.2. By Corollary 11.4, it suffices to evaluate the sum in (11.18). However, by (6.2), this sum is just the specialization (11.17) of the product $\mathcal{E}$ in (6.2). The product of the first $2m + 2n$ factors of $\mathcal{E}$ specializes to

$$(11.19) \qquad \prod_{k=1}^{m}\left(\frac{2iR_k}{\sqrt{3}}\right)^2 \prod_{j=1}^{n}\left(\frac{2iR'_j}{\sqrt{3}}\right)^2 = \frac{(-4)^{m+n}}{3^{m+n}}\prod_{k=1}^{m}R_k^2 \prod_{j=1}^{n}(R'_j)^2.$$

The product of the next $4\binom{m}{2}$ factors of $\mathcal{E}$ specializes to

$$\prod_{1\le k<j\le m}\left\{\left((v_j - v_k) + \left(\frac{iR_j}{\sqrt{3}} - \frac{iR_k}{\sqrt{3}}\right)\right)\left((v_j - v_k) + \left(\frac{iR_j}{\sqrt{3}} + \frac{iR_k}{\sqrt{3}}\right)\right)\right.$$

$$\left. \times \left((v_j - v_k) + \left(-\frac{iR_j}{\sqrt{3}} - \frac{iR_k}{\sqrt{3}}\right)\right)\left((v_j - v_k) + \left(-\frac{iR_j}{\sqrt{3}} + \frac{iR_k}{\sqrt{3}}\right)\right)\right\}$$

$$(11.20)$$

$$= \prod_{1\le k<j\le m}\left[(v_j - v_k)^2 + \frac{1}{3}(R_j - R_k)^2\right]\left[(v_j - v_k)^2 + \frac{1}{3}(R_j + R_k)^2\right],$$

while the product of the last $4\binom{n}{2}$ factors of $\mathcal{E}$ specializes to

$$\prod_{1\le k<j\le n}\left\{\left((v'_j - v'_k) + \left(\frac{iR'_j}{\sqrt{3}} - \frac{iR'_k}{\sqrt{3}}\right)\right)\left((v'_j - v'_k) + \left(\frac{iR'_j}{\sqrt{3}} + \frac{iR'_k}{\sqrt{3}}\right)\right)\right.$$

$$\left. \times \left((v'_j - v'_k) + \left(-\frac{iR'_j}{\sqrt{3}} - \frac{iR'_k}{\sqrt{3}}\right)\right)\left((v'_j - v'_k) + \left(-\frac{iR'_j}{\sqrt{3}} + \frac{iR'_k}{\sqrt{3}}\right)\right)\right\}$$

$$(11.21)$$

$$= \prod_{1\le k<j\le n}\left[(v'_j - v'_k)^2 + \frac{1}{3}(R'_j - R'_k)^2\right]\left[(v'_j - v'_k)^2 + \frac{1}{3}(R'_j + R'_k)^2\right].$$

In the expression Y given by (11.15) write, using (2.3),

$$\sqrt{q_k^2 + \frac{1}{3}} = \frac{1}{R_k}\sqrt{(q_k R_k)^2 + \frac{1}{3}R_k^2} = \frac{1}{R_k}\sqrt{(v_k - c_k)^2 + \frac{1}{3}R_k^2},$$

$$\sqrt{q'^2_j + \frac{1}{3}} = \frac{1}{R'_j}\sqrt{(q'_j R'_j)^2 + \frac{1}{3}(R'_j)^2} = \frac{1}{R'_j}\sqrt{(v'_j - c'_j)^2 + \frac{1}{3}(R'_j)^2},$$

and

$$\left[(q_k A_k + q'_j B_j)^2 + \frac{1}{3}(A_k \pm B_j)^2\right] = \frac{1}{R^2}\left[(q_k A_k R + q'_j B_j R)^2 + \frac{1}{3}(A_k R \pm B_j R)^2\right]$$

$$= \frac{1}{R^2}\left[(v_k + v'_j - c_k - c'_j)^2 + \frac{1}{3}(R_k \pm R'_j)^2\right].$$

Substituting the resulting expression for Y and formula (11.13) for X into (11.18), we obtain by Corollary 11.4 and (11.19)–(11.21) that

$$\omega_b \begin{pmatrix} R_1 & \dots & R_m & ; & R_1' & \dots & R_n' \\ v_1 & & v_m & & v_1' & & v_n' \end{pmatrix} = \chi_{2m,2n} \prod_{k=1}^{m} R_k \prod_{j=1}^{n} R_j'(R_j' - 1/2)(R_j' + 1/2)$$

$$\times \frac{2^{m-n}3^n}{\pi^{m+n}} \frac{1}{\prod_{k=1}^{m} R_k^2 \prod_{j=1}^{n} (R_j')^4} \frac{\prod_{j=1}^{n} \sqrt{(v_j' - c_j')^2 + \frac{1}{3}(R_j')^2}}{\prod_{k=1}^{m} \sqrt{(v_k - c_k)^2 + \frac{1}{3} R_k^2}}$$

$$\times \frac{1}{\prod_{k=1}^{m} \prod_{j=1}^{n} [(v_k + v_j' - c_k - c_j')^2 + \frac{1}{3}(R_k - R_j')^2]}$$

$$\times \frac{1}{\prod_{k=1}^{m} \prod_{j=1}^{n} [(v_k + v_j' - c_k - c_j')^2 + \frac{1}{3}(R_k + R_j')^2]}$$

$$\times \Bigg| \frac{(-4)^{m+n}}{3^{m+n}} \prod_{k=1}^{m} R_k^2 \prod_{j=1}^{n} (R_j')^2$$

$$\times \prod_{1 \le k < j \le m} \left[(v_j - v_k)^2 + \frac{1}{3}(R_j - R_k)^2 \right] \left[(v_j - v_k)^2 + \frac{1}{3}(R_j + R_k)^2 \right]$$

$$\times \prod_{1 \le k < j \le n} \left[(v_j' - v_k')^2 + \frac{1}{3}(R_j' - R_k')^2 \right] \left[(v_j' - v_k')^2 + \frac{1}{3}(R_j' + R_k')^2 \right] \Bigg|$$

$$+ O(R^{2m^2 + 2n^2 - 4mn - 2m - 1}).$$

(11.22)

By (2.3), the parameters $R_1, \dots, R_m$, $v_1, \dots, v_m$ and $R_1', \dots, R_n'$, $v_1', \dots, v_n'$ approach infinity as $R \to \infty$, while $c_1, \dots, c_m$ and $c_1', \dots, c_n'$ are constant. It follows from this that the difference between the product on the right hand side of (11.22) and what it becomes when one omits the additive constants $\pm 1/2$ on its first line and the constants c_k and c_j' is $O(R^{2m^2 + 2n^2 - 4mn - 2m - 1})$. The latter is readily brought to the form given by (2.9) and (2.10). By the assumption on the distinctness of the pairs (A_k, q_k), $k = 1, \dots, m$, and (B_j, q_j'), $j = 1, \dots, n$ in (2.3), the expression on the right hand side of (2.9) has degree $2m^2 + 2n^2 - 4mn - 2m$ in R. Therefore, (2.9) does indeed give the asymptotics of ω_b. $\quad\square$

Another simple product formula
for correlations along the boundary

By Proposition 3.2, the asymptotics of the joint correlation at the center ω will follow provided we also work out the asymptotics of the boundary-influenced correlation $\bar{\omega}_b$ defined by (3.3). We present this in Section 13. But first we need an analog of Proposition 4.1 corresponding to the regions E_N used to define $\bar{\omega}_b$.

Let E be the region determined by the common outside boundary of the regions $E_N \begin{pmatrix} R_1 & \dots & R_m & R'_1 & \dots & R'_n \\ v_1 & & v_m & v'_1 & & v'_n \end{pmatrix}$ defined in Section 3, for fixed m, n and N. Then E is the half-hexagonal lattice region with four straight sides—the southern side of length $N + 2n + 1$, southeastern of length $2N + 4m - 1$, northeastern of length $2N + 4n$, and northern of length $N + 2m$—followed by $N + 2n$ descending zig-zags to the lattice point O, two extra unit steps southeast of O followed by one step west, and $N + 2m - 1$ more descending zig-zags to close up the boundary (an example can be seen in Figure 12.1). In addition, the $N + 2m - 1$ dimer positions weighted $1/2$ in the regions E_N are also weighted so in E.

As in the case of the regions W of Section 3, the vertical, jagged boundary of E can be viewed as consisting of bumps—in the present case, pairs of adjacent lattice segments forming an angle that opens to the *east*: $N + 2m - 1$ bumps below O, and $N + 2n$ above O. Label the former by $0, 1, \dots, N + 2m - 2$ and the latter by $0, 1, \dots, N + 2n - 1$, both labelings starting with the bumps closest to O and then moving successively outwards.

In the description of E the parameters m and n always appear with even coefficients. We re-denote, as in Section 3, $2m$ by m and $2n$ by n, for notational simplicity. Therefore we consider the four straight sides of E to have lengths $N + n$, $2N + 2m - 1$, $2N + 2n$ and $N + m$, while the number of bumps below and above O is $N + m - 1$ and $N + n$, respectively.

We allow removal of any bump exactly like in Section 3: above O, place an up-pointing quadromer across it and discard the three monomers of E it covers; below O, use down-pointing quadromers.

Define $E_N[k_1, \dots, k_m; l_1, \dots, l_n]$ to be the region obtained from E by removing the bumps below O with labels $0 \le k_1 < k_2 < \dots < k_m \le N + m - 2$, and the bumps above O with labels $0 \le l_1 < l_2 < \dots < l_n \le N + n - 1$. Figure 12.1 shows $E_2[1, 2, 3, 4; 1, 2, 3, 4, 5, 6]$.

The correlations of the removed bumps on the boundary of the regions E_N turn out to be *exactly* the same as the corresponding correlations we found in Section 3 for the regions W_N.

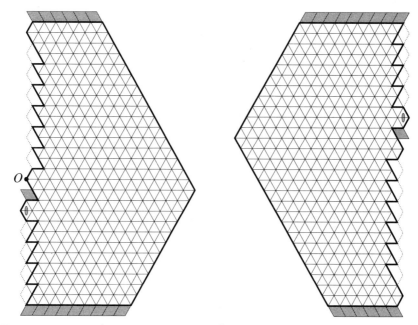

FIGURE 12.1. $E_2[1, 2, 3, 4; 1, 2, 3, 4, 5, 6]$.   FIGURE 12.2. Region denoted $R_{[2,3,4,5,6,7][2,3,4,5]}(6)$ in the paper in Part B.

PROPOSITION 12.1. *For $m, n \geq 0$ and fixed integers $0 \leq k_1 < k_2 < \cdots < k_m$ and $0 \leq l_1 < l_2 < \cdots < l_n$ we have*

(12.1)
$$\lim_{N \to \infty} \frac{\mathrm{M}\left(E_N[k_1, \ldots, k_m; l_1, \ldots, l_n]\right)}{\mathrm{M}\left(E_N[0, \ldots, m-1; 0, \ldots, n-1]\right)} =$$

$$= \chi_{m,n} \prod_{i=1}^{m} \frac{(3/2)_{k_i}}{(2)_{k_i}} \prod_{i=1}^{n} \frac{(3/2)_{l_i}}{(1)_{l_i}} \frac{\displaystyle\prod_{1 \leq i < j \leq m} (k_j - k_i) \prod_{1 \leq i < j \leq n} (l_j - l_i)}{\displaystyle\prod_{i=1}^{m} \prod_{i=1}^{n} (k_i + l_j + 2)},$$

where $\chi_{m,n}$ is given by (4.3).

PROOF. We proceed in analogy to the proof of Proposition 4.1. The results of the paper in Part B of this Memoir provide an explicit formula for

$$\mathrm{M}(E_N[k_1, \ldots, k_m; l_1, \ldots, l_n])$$

as well. Indeed, in the notation of that paper we have

(12.2)
$$E_N[k_1, \ldots, k_m; l_1, \ldots, l_n] =$$

$$R_{[1,\ldots,N+n] \setminus [l_1+1,\ldots,l_n+1],[1,\ldots,N-1+m] \setminus [k_1+1,\ldots,k_m+1]}(N+m)$$

(this is illustrated by Figure 12.2, which is just the 180° rotation of Figure 12.1).

Proposition 2.1 of the paper in Part B together with formulas (1.1), (1.3) and (1.5) from that paper provide an explicit formula for $\mathrm{M}(R_{\mathbf{p},\mathbf{q}}(x))$, for any pair of lists $\mathbf{p} = [p_1, \ldots, p_s]$, $1 \le p_1 < \cdots < p_s$ and $\mathbf{q} = [q_1, \ldots, q_t]$, $1 \le q_1 < \cdots < q_t$, and any nonnegative integer $x \ge q_t - p_s - t + s - 1$.

Written as a constant times a monic polynomial in x, this formula takes the form

$$(12.3) \qquad \qquad \mathrm{M}\left(R_{\mathbf{p},\mathbf{q}}(x)\right) = c_{\mathbf{p},\mathbf{q}} G_{\mathbf{p},\mathbf{q}}(x),$$

where by formula (1.3) of Part B we have
$$(12.4)$$
$$c_{\mathbf{p},\mathbf{q}} = 2^{\binom{t-s}{2}-s} \prod_{i=1}^{s} \frac{1}{(2p_i)!} \prod_{i=1}^{t} \frac{1}{(2q_i - 1)!} \frac{\prod_{1 \le i < j \le s}(p_j - p_i) \prod_{1 \le i < j \le t}(q_j - q_i)}{\prod_{i=1}^{s} \prod_{i=1}^{t}(p_i + q_j)}.$$

Furthermore, it is straightforward to check, using formulas (1.1) and (1.5) of Part B, that the polynomials $G_{\mathbf{p},\mathbf{q}}(x)$ satisfy

$$(12.5)$$
$$\frac{G_{\mathbf{p}^{|i\rangle},\mathbf{q}}(x)}{G_{\mathbf{p},\mathbf{q}}(x)} = (x - p_i + p_s)(x + p_i + p_s - s + t + 2), \qquad \text{for } 1 \le i < s$$

$$(12.6)$$
$$\frac{G_{\mathbf{p},\mathbf{q}^{|i\rangle}}(x)}{G_{\mathbf{p},\mathbf{q}}(x)} = (x + q_i + p_s + 1)(x - q_i + p_s - s + t + 1), \qquad \text{for } 1 \le i \le t$$

(as in Section 4, $\mathbf{p}^{|i\rangle}$ denotes the list obtained from $\mathbf{p}$ by increasing its i-th element by 1, and is defined only if $l_{i+1} - l_i \ge 2$).

Consider first the limit

$$(12.7) \qquad \lim_{N \to \infty} \frac{\mathrm{M}\left(E_N[k_1, \ldots, k_{i-1}, k_i + 1, k_{i+1}, \ldots, k_m; l_1, \ldots, l_n]\right)}{\mathrm{M}\left(E_N[k_1, \ldots, k_{i-1}, k_i, k_{i+1}, \ldots, k_m; l_1, \ldots, l_n]\right)},$$

for $k_i + 1 < k_{i+1}$.

Use (12.2) to view the regions involved in this fraction as $R_{\mathbf{p},\mathbf{q}}(x)$'s. Comparing formulas (12.4) and (4.6), and using $\binom{s-t}{2} - t = \binom{t-s}{2} - s$, one sees that

$$c_{\mathbf{p},\mathbf{q}} = \bar{c}_{\mathbf{q},\mathbf{p}}.$$

For the lists on the right hand side of (12.2) this implies

$$(12.8) \qquad \begin{aligned} & c_{[1,\ldots,N+n] \setminus [l_1+1,\ldots,l_n+1],[1,\ldots,N-1+m] \setminus [k_1+1,\ldots,k_m+1]} \\ & = \bar{c}_{[1,\ldots,N-1+m] \setminus [k_1+1,\ldots,k_m+1],[1,\ldots,N+n] \setminus [l_1+1,\ldots,l_n+1]}. \end{aligned}$$

Therefore, the contribution to the fraction in (12.7) coming (via (12.2)) from the $c_{\mathbf{p},\mathbf{q}}$-parts of (12.3) follows from our work in Section 4. The only difference from that case is that in the $\bar{c}_{\mathbf{p},\mathbf{q}}$ of (12.8) the largest element of $\mathbf{p}$ is now $(N-1)+m$, as opposed to $N + m$. However, substitution of N by $N - 1$ in (4.10) makes no difference in the limit $N \to \infty$, so the contribution of the $c_{\mathbf{p},\mathbf{q}}$-parts to the limit (12.7) is, asymptotically for $N \to \infty$, precisely the same as (4.10).

On the other hand, one readily sees by (12.2) and (12.6) that the contribution of the $G_{\mathbf{p},\mathbf{q}}(x)$-parts to the fraction in (12.7) is

$$\frac{1}{(2N + m + n + k_i + 2)(2N + m + n - k_i - 1)},$$

a fraction whose asymptotics as $N \to \infty$ is clearly the same as that of (4.11).

It follows from the above two paragraphs that the limit (12.7) is equal to the product on the right hand side of (4.12), so it has exactly the same expression as in the case of the regions W_N treated in Section 4.

A similar analysis shows that

$$
\lim_{N \to \infty} \frac{M\left(E_N[k_1, \ldots, k_m; l_1, \ldots, l_{i-1}, l_i + 1, l_{i+1}, \ldots, l_n]\right)}{M\left(E_N[k_1, \ldots, k_m; l_1, \ldots, l_{i-1}, l_i, l_{i+1}, \ldots, l_n]\right)}
$$
$$
= \lim_{N \to \infty} \frac{M\left(W_N[k_1, \ldots, k_m; l_1, \ldots, l_{i-1}, l_i + 1, l_{i+1}, \ldots, l_n]\right)}{M\left(W_N[k_1, \ldots, k_m; l_1, \ldots, l_{i-1}, l_i, l_{i+1}, \ldots, l_n]\right)},
$$

so that also decrementing a single element from the second index list has exactly the same effect for the regions E_N as for the regions W_N of Section 4. Formula (12.1) follows therefore by Proposition 4.1. $\square$

The asymptotics of $\bar{\omega}_b$. Proof of Theorem 2.1

Our reasoning from Section 5 applies with no change to the regions E_N defined in Section 3. Each missing quadromer can be placed back at the expense of performing a Laplace expansion in the corresponding Gessel-Viennot determinant. The resulting 2×2 cofactors have precisely the forms (5.5) and (5.13). The analog of (5.19) we get is

$$M\left(E_N\begin{pmatrix} R_1 & \cdots & R_m \\ v_1 & & v_m \end{pmatrix}; \begin{matrix} R_1' & \cdots & R_n' \\ v_1' & & v_n' \end{matrix}\right) = 2^{m+n} \prod_{i=1}^{m} R_i(R_i - 1/2)(R_i + 1/2) \prod_{i=1}^{n} R_i'$$

$$\times \left| \sum_{0 \le a_1 < b_1 \le R_1} \cdots \sum_{0 \le a_m < b_m \le R_m} \sum_{0 \le c_1 < d_1 \le R_1'} \cdots \sum_{0 \le c_n < d_n \le R_n'} (-1)^{\sum_{i=1}^{m}(a_i+b_i)} \right.$$

$$\times (-1)^{\sum_{i=1}^{n}(c_i+d_i)}$$

$$\times \prod_{i=1}^{m} \frac{(b_i - a_i)(R_i + a_i - 1)!\,(R_i + b_i - 1)!}{(2a_i + 1)!\,(R_i - a_i)!\,(2b_i + 1)!\,(R_i - b_i)!}$$

$$\times \prod_{i=1}^{n} \frac{(d_i - c_i)(R_i' + c_i - 1)!\,(R_i' + d_i - 1)!}{(2c_i)!\,(R_i' - c_i)!\,(2d_i)!\,(R_i' - d_i)!}$$

$$\times \operatorname{sgn}(v_1' + c_1, v_1' + d_1, \ldots, v_n' + c_n, v_n' + d_n)$$

$$\times \operatorname{sgn}(v_1 + a_1, v_1 + b_1, \ldots, v_m + a_m, v_m + b_m)$$

$$\times M(E_N[\{v_1 + a_1, \ldots, v_m + b_m\}_<; \{v_1' + c_1, \ldots, v_n' + d_n\}_<]) \bigg|,$$

(13.1)

where the summation extends only over those summation indices for which both lists of arguments in the region on the right hand side have distinct elements, and sgn denotes the sign of such a list when regarded as a permutation.

After removing the forced dimers, the normalizing region at the denominator of (3.3),

(13.2) $$E_N\begin{pmatrix} 1 & 3 & \cdots & 2m-1 \\ 0 & 0 & & 0 \end{pmatrix}; \begin{matrix} 1 & 3 & \cdots & 2n-1 \\ 0 & 0 & & 0 \end{matrix}\end{pmatrix},$$

is readily seen to be precisely $E_N[0, 1, \ldots, 2m-1; 0, 1, \ldots, 2n-1]$.

Moreover, by the sameness of the right hand sides of (12.1) and (4.2), (5.21) implies

$$
\lim_{N\to\infty} \frac{\mathrm{M}\left(E_N[\{v_1+a_1,\ldots,v_m+b_m\}_<;\{v'_1+c_1,\ldots,v'_n+d_n\}_<]\right)}{\mathrm{M}\left(E_N[0,\ldots,2m-1;0,\ldots,2n-1]\right)} = \chi_{2m,2n}
$$

$$
\times\,\mathrm{sgn}(v_1+a_1,v_1+b_1,\ldots,v_m+a_m,v_m+b_m)
$$

$$
\times\,\mathrm{sgn}(v'_1+c_1,v'_1+d_1,\ldots,v'_n+c_n,v'_n+d_n)
$$

$$
\times\prod_{i=1}^m \frac{(3/2)_{v_i+a_i}\,(3/2)_{v_i+b_i}}{(2)_{v_i+a_i}\,(2)_{v_i+b_i}}\prod_{i=1}^n \frac{(3/2)_{v'_i+c_i}\,(3/2)_{v'_i+d_i}}{(1)_{v'_i+c_i}\,(1)_{v'_i+d_i}}
$$

$$
\times\prod_{1\le i<j\le m}(v_j-v_i+a_j-a_i)(v_j-v_i+a_j-b_i)
$$

$$
\times\,(v_j-v_i+b_j-a_i)(v_j-v_i+b_j-b_i)
$$

$$
\times\prod_{1\le i<j\le n}(v'_j-v'_i+c_j-c_i)(v'_j-v'_i+c_j-d_i)
$$

$$
\times\,(v'_j-v'_i+d_j-c_i)(v'_j-v'_i+d_j-d_i)
$$

$$
\times\,\frac{\prod_{i=1}^m(a_i-b_i)\prod_{i=1}^n(c_i-d_i)}{\prod_{i=1}^m\prod_{j=1}^n(u_{ij}+a_i+c_j)(u_{ij}+a_i+d_j)(u_{ij}+b_i+c_j)(u_{ij}+b_i+d_j)},
$$

(13.3)

where χ is given by (4.3) and $u_{ij}=v_i+v'_j+2$, $i=1,\ldots,m$, $j=1,\ldots,n$.

Divide (13.1) by the matching generation function of the region (13.2) and let $N\to\infty$. By (3.3) and (13.3) we obtain

$$
\bar{\omega}_b\begin{pmatrix} R_1 & \cdots & R_m \\ v_1 & & v_m \end{pmatrix}; \begin{matrix} R'_1 & \cdots & R'_n \\ v'_1 & & v'_n \end{matrix} = 2^{m+n}\chi_{m,n}\prod_{i=1}^m R_i(R_i-1/2)(R_i+1/2)\prod_{i=1}^n R'_i
$$

$$
\times\Bigg|\sum_{0\le a_1<b_1\le R_1}\cdots\sum_{0\le a_m<b_m\le R_m}\sum_{0\le c_1<d_1\le R'_1}\cdots\sum_{0\le c_n<d_n\le R'_n}(-1)^{\sum_{i=1}^m(a_i+b_i)}
$$

$$
\times\,(-1)^{\sum_{i=1}^n(c_i+d_i)}
$$

$$
\times\prod_{i=1}^m \frac{(R_i+a_i-1)!\,(R_i+b_i-1)!}{(2a_i+1)!\,(R_i-a_i)!\,(2b_i+1)!\,(R_i-b_i)!}
$$

$$
\times\prod_{i=1}^n \frac{(R_i+c_i-1)!\,(R_i+d_i-1)!}{(2c_i)!\,(R_i-c_i)!\,(2d_i)!\,(R_i-d_i)!}
$$

$$
\times\prod_{i=1}^m \frac{(3/2)_{v_i+a_i}\,(3/2)_{v_i+b_i}}{(2)_{v_i+a_i}\,(2)_{v_i+b_i}}\prod_{i=1}^n \frac{(3/2)_{v'_i+c_i}\,(3/2)_{v'_i+d_i}}{(1)_{v'_i+c_i}\,(1)_{v'_i+d_i}}
$$

$$\times \prod_{1 \le i < j \le m} (v_j - v_i + a_j - a_i)(v_j - v_i + a_j - b_i)$$

$$\times (v_j - v_i + b_j - a_i)(v_j - v_i + b_j - b_i)$$

$$\times \prod_{1 \le i < j \le n} (v'_j - v'_i + c_j - c_i)(v'_j - v'_i + c_j - d_i)$$

$$\times (v'_j - v'_i + d_j - c_i)(v'_j - v'_i + d_j - d_i)$$

$$\times \frac{\prod_{i=1}^m (a_i - b_i)^2 \prod_{i=1}^n (c_i - d_i)^2}{\prod_{i=1}^m \prod_{j=1}^n (u_{ij} + a_i + c_j)(u_{ij} + a_i + d_j)(u_{ij} + b_i + c_j)(u_{ij} + b_i + d_j)} \Bigg|,$$

(13.4)

where the summation range is restricted to those summation variables for which $v_1 + a_1, v_1 + b_1, \ldots, v_m + a_m, v_m + b_m$, as well as $v'_1 + c_1, v'_1 + d_1, \ldots, v'_n + c_n, v'_n + d_n$, are distinct.

By the argument in the last paragraph of the proof of Lemma 5.1, the restrictions $a_i < b_i$, $i = 1, \ldots, m$, and $c_i < d_i$, $i = 1, \ldots, n$ can be dropped at the expense of a multiplicative factor of $1/2^{m+n}$. We obtain the following result.

LEMMA 13.1. *For fixed* $R_1, \ldots, R_m, R'_1, \ldots, R'_n \ge 1$ *and* $v_1, \ldots, v_m$, $v'_1, \ldots, v'_n \ge 0$ *we have*

$$\bar{\omega}_b \begin{pmatrix} R_1 & \ldots & R_m \\ v_1 & & v_m \end{pmatrix} ; \begin{matrix} R'_1 & \ldots & R'_n \\ v'_1 & & v'_n \end{matrix} = \chi_{2m,2n} \prod_{i=1}^m R_i(R_i - 1/2)(R_i + 1/2) \prod_{i=1}^n R'_i$$

$$\times \Bigg| \sum_{a_1,b_1=0}^{R_1} \cdots \sum_{a_m,b_m=0}^{R_m} \sum_{c_1,d_1=0}^{R'_1} \cdots \sum_{c_n,d_n=0}^{R'_n} (-1)^{\sum_{i=1}^m (a_i + b_i) + \sum_{i=1}^n (c_i + d_i)}$$

$$\times \prod_{i=1}^m \frac{(R_i + a_i - 1)! \, (R_i + b_i - 1)!}{(2a_i + 1)! \, (R_i - a_i)! \, (2b_i + 1)! \, (R_i - b_i)!} \frac{(3/2)_{v_i + a_i} \, (3/2)_{v_i + b_i}}{(2)_{v_i + a_i} \, (2)_{v_i + b_i}}$$

$$\times \prod_{i=1}^n \frac{(R_i + c_i - 1)! \, (R_i + d_i - 1)!}{(2c_i)! \, (R_i - c_i)! \, (2d_i)! \, (R_i - d_i)!} \frac{(3/2)_{v'_i + c_i} \, (3/2)_{v'_i + d_i}}{(1)_{v'_i + c_i} \, (1)_{v'_i + d_i}}$$

$$\times \prod_{1 \le i < j \le m} (v_j - v_i + a_j - a_i)(v_j - v_i + a_j - b_i)(v_j - v_i + b_j - a_i)(v_j - v_i + b_j - b_i)$$

$$\times \prod_{1 \le i < j \le n} (v'_j - v'_i + c_j - c_i)(v'_j - v'_i + c_j - d_i)(v'_j - v'_i + d_j - c_i)(v'_j - v'_i + d_j - d_i)$$

$$\times \frac{\prod_{i=1}^m (a_i - b_i)^2 \prod_{i=1}^n (c_i - d_i)^2}{\prod_{i=1}^m \prod_{j=1}^n (u_{ij} + a_i + c_j)(u_{ij} + a_i + d_j)(u_{ij} + b_i + c_j)(u_{ij} + b_i + d_j)} \Bigg|,$$

(13.5)

where

$$u_{ij} = v_i + v'_j + 2$$

for $i = 1, \ldots, m$, $j = 1, \ldots, n$, and χ is given by (4.3).

Note that expressions (13.5) and (5.1) are nearly identical. There are only two differences: first, the roles of R_i and R'_i are interchanged in the products on the first lines of their right hand sides; and second, in the denominators on the third and fourth lines of (13.5) one has the expressions $(2a_i + 1)!$, $(2b_i + 1)!$, $(2c_i)!$ and $(2d_i)!$, while in the corresponding positions in (5.1) one has $(2a_i)!$, $(2b_i)!$, $(2c_i + 1)!$ and $(2d_i + 1)!$, respectively.

The fact that these are the only differences allows us to obtain the asymptotics of $\bar{\omega}_b$ from the asymptotics of ω_b worked out in Sections 6–12.

The arguments of Section 6 apply equally well to yield an expression for $\bar{\omega}_b$ analogous to the expression (6.5) for ω_b. By the the first two formulas in the proof of Proposition 6.1, besides interchanging the roles of R_i and R'_i in the products on the first line of (5.1), the only effect of the difference between (13.5) and (5.1) is that the fractions $1/2$ and $3/2$ at the denominators of (6.7) and (6.8) swap places for the sums $\bar{T}^{(n)}$ and $\bar{T}'^{(n)}$—the analogs of $T^{(n)}$ and $T'^{(n)}$. More precisely, define

$$(13.6) \qquad \bar{T}^{(n)}(R, v; x) := \frac{1}{R} \sum_{a=0}^{R} \frac{(-R)_a \, (R)_a \, (3/2)_{v+a}}{(1)_a \, (3/2)_a \, (2)_{v+a}} \left(\frac{x}{4}\right)^a a^n$$

$$(13.7) \qquad \bar{T}'^{(n)}(R, v; x) := \frac{1}{R} \sum_{c=0}^{R} \frac{(-R)_c \, (R)_c \, (3/2)_{v+c}}{(1)_c \, (1/2)_c \, (1)_{v+c}} \left(\frac{x}{4}\right)^c c^n.$$

Then by the arguments that proved Proposition 6.1 we obtain the following result.

PROPOSITION 13.2. *The boundary-influenced correlation $\bar{\omega}_b$ can be written as*

$$\bar{\omega}_b \begin{pmatrix} R_1 & \ldots & R_m \\ v_1 & & v_m \end{pmatrix} \begin{matrix} R'_1 & \ldots & R'_n \\ v'_1 & & v'_n \end{matrix} = \chi_{2m,2n} \prod_{i=1}^{m} R_i(R_i - 1/2)(R_i + 1/2) \prod_{i=1}^{n} R'_i$$

$$(13.8)$$

$$\times \left| \sum_{C \in \mathcal{C}} e(C) \bar{M}_{\alpha_1(C), \beta_1(C), \ldots, \alpha_m(C), \beta_m(C); \gamma_1(C), \delta_1(C), \ldots, \gamma_n(C), \delta_n(C)} \right|,$$

where χ is given by (4.3), the collection $\mathcal{C}$ and $e(C)$, $\alpha_i(C)$, $\beta_i(C)$, $\gamma_j(C)$, $\delta_j(C)$ are as in (6.2), and the "moment" $\bar{M}_{\alpha_1, \beta_1, \ldots, \alpha_m, \beta_m; \gamma_1, \delta_1, \ldots, \gamma_n, \delta_n}$ equals the 4mn-fold *integral*

$$\bar{M}_{\alpha_1,\beta_1,\ldots,\alpha_m,\beta_m;\gamma_1,\delta_1,\ldots,\gamma_n,\delta_n} = \int_0^1 \cdots \int_0^1 \prod_{i=1}^m \prod_{j=1}^n (x_{ij} y_{ij} z_{ij} w_{ij})^{v_i + v'_j + 1}$$

$$\times \bar{T}^{(\alpha_1)}(R_1, v_1; \prod_{j=1}^n x_{1j} y_{1j}) \, \bar{T}^{(\beta_1)}(R_1, v_1; \prod_{j=1}^n z_{1j} w_{1j}) \cdots$$

$$\cdots \bar{T}^{(\alpha_m)}(R_m, v_m; \prod_{j=1}^n x_{mj} y_{mj}) \, \bar{T}^{(\beta_m)}(R_m, v_m; \prod_{j=1}^n z_{mj} w_{mj})$$

$$\times \bar{T}'^{(\gamma_1)}(R'_1, v'_1; \prod_{i=1}^m x_{i1} z_{i1}) \, \bar{T}'^{(\delta_1)}(R'_1, v'_1; \prod_{i=1}^m y_{i1} w_{i1}) \cdots$$

(13.9)

$$\cdots \bar{T}'^{(\gamma_n)}(R'_n, v'_n; \prod_{i=1}^m x_{in} z_{in}) \, \bar{T}'^{(\delta_n)}(R'_n, v'_n; \prod_{i=1}^m y_{in} w_{in}) \, dx_{11} \cdots dw_{mn},$$

with $\bar{T}^{(n)}(R, v; x)$ and $\bar{T}'^{(n)}(R, v; x)$ defined by (13.6) and (13.7).

Furthermore, the calculations that proved Lemma 6.2 show that

$$\bar{T}^{(n)}(R, v; x) =$$
$$\frac{1}{R} \frac{(3/2)_v}{(2)_v} \sum_{k=0}^n f_k \frac{(-R)_k \, (R)_k \, (v + 3/2)_k}{(3/2)_k \, (v + 2)_k} \left(\frac{x}{4}\right)^k {}_3F_2\left[\begin{matrix} -R + k, \, R + k, \, \frac{3}{2} + v + k \\ \frac{3}{2} + k, \, 2 + v + k \end{matrix}; \frac{x}{4}\right]$$

(13.10)

$$\bar{T}'^{(n)}(R, v; x) =$$
$$\frac{1}{R} \frac{(3/2)_v}{(1)_v} \sum_{k=0}^n f_k \frac{(-R)_k \, (R)_k \, (v + 3/2)_k}{(1/2)_k \, (v + 1)_k} \left(\frac{x}{4}\right)^k {}_3F_2\left[\begin{matrix} -R + k, \, R + k, \, \frac{3}{2} + v + k \\ \frac{1}{2} + k, \, 1 + v + k \end{matrix}; \frac{x}{4}\right],$$

(13.11)

where the f_k's are as in (6.9) (in particular $f_n = 1$). These expressions are nearly the same as (6.10) and (6.11): the only difference is that, compared to $T^{(n)}$ and $T'^{(n)}$, the fractions $1/2$ and $3/2$ at the denominators (including the denominator parameters of the ${}_3F_2$'s) swap places in the "barred" versions $\bar{T}^{(n)}$ and $\bar{T}'^{(n)}$.

Using (7.8) to write the ${}_3F_2$'s of (13.10) as integrals of ${}_2F_1$'s, the derivation rule for ${}_2F_1$'s displayed after (7.9), and the ${}_2F_1$ evaluation (7.21), we obtain after simplifications

$$\bar{T}^{(n)}(R, qR + c; x) = \frac{2}{\pi R} \sum_{k=0}^{n} f_k \int_0^1 t^{qR+c+k+1/2}(1 - t)^{-1/2}$$

$$\times \frac{d^k}{dt^k} \left\{ \frac{2R}{4R^2 - 1} \sqrt{\frac{4 - xt}{xt}} \sin\left[R \arccos\left(1 - \frac{xt}{2}\right)\right]\right.$$

(13.12)

$$\left. - \frac{1}{4R^2 - 1} \cos\left[R \arccos\left(1 - \frac{xt}{2}\right)\right]\right\} dt.$$

To obtain a similar expression for $\bar{T}'^{(n)}$ we need first an analog of (7.16). One readily sees that the calculation that proved (7.16) also shows that

$$_3F_2\left[\begin{matrix} -R+k, \, R+k, \, \frac{3}{2}+v+k \\ \frac{1}{2}+k, \, 1+v+k \end{matrix}; \frac{x}{4}\right] = {}_3F_2\left[\begin{matrix} -R+k, \, R+k, \, \frac{1}{2}+v+k \\ \frac{1}{2}+k, \, 1+v+k \end{matrix}; \frac{x}{4}\right]$$

$$+ \frac{x}{4} \frac{(-R+k)(R+k)}{(1/2+k)(1+v+k)} {}_3F_2\left[\begin{matrix} -R+k+1, \, R+k+1, \, \frac{3}{2}+v+k \\ \frac{3}{2}+k, \, 2+v+k \end{matrix}; \frac{x}{4}\right].$$

Express the $_3F_2$'s of (13.11) using the above identity. Use (7.8) to write the resulting $_3F_2$'s as integrals of $_2F_1$'s. In turn, express the latter, using the derivation rule for $_2F_1$'s displayed after (7.9), in terms of the $_2F_1$ evaluation (7.11). We obtain after simplifications

$$\bar{T}'^{(n)}(R, qR + c; x) =$$

$$\frac{2}{\pi R} \sum_{k=0}^{n} f_k(qR + c + k + 1/2) \int_0^1 t^{qR} \frac{t^{k+c-1/2}}{(1 - t)^{1/2}} \frac{d^k}{dt^k} \cos\left[R \arccos\left(1 - \frac{xt}{2}\right)\right] dt$$

$$+ \frac{2}{\pi R} \sum_{k=0}^{n} f_k \int_0^1 t^{qR} \frac{t^{k+c+1/2}}{(1 - t)^{1/2}} \frac{d^{k+1}}{dt^{k+1}} \cos\left[R \arccos\left(1 - \frac{xt}{2}\right)\right] dt.$$

(13.13)

The asymptotics of the $\bar{T}^{(n)}$'s and $\bar{T}'^{(n)}$'s can be obtained using the same approach we employed in Section 7 for the $T^{(n)}$'s and $T'^{(n)}$'s. Indeed, by Lemma 7.4, when (7.23) is substituted in (13.12) the omitted terms in (7.23) give rise to integrals like in Proposition 7.2. By the argument in the proof of Proposition 7.1, we deduce that the asymptotics of $\bar{T}^{(n)}(R, qR+c; x)$ is generated by the main term on the right hand side of (7.23) for $k = n$. More precisely, we obtain

$$\left| \bar{T}^{(n)}(R, qR + c; x) - \frac{1}{\pi R^2} \int_0^1 t^{qR} \frac{t^{n+c+1/2}}{(1 - t)^{1/2}} \sqrt{\frac{4 - xt}{xt}} \left(R\sqrt{\frac{x}{4t - xt^2}}\right)^n \right.$$

(13.14)

$$\left. \times \cos\left[R \arccos\left(1 - \frac{xt}{2}\right) + \frac{(n - 1)\pi}{2}\right] dt \right| \leq \bar{M}_0 R^{n-7/2},$$

for $R \geq \bar{R}_0$, with $\bar{R}_0$ and $\bar{M}_0$ independent of $x \in [0, 1]$. Approximating the integral in (13.14) by Proposition 7.1 we deduce the first part of the following result.

PROPOSITION 13.3. *Let $q > 0$ be a fixed rational number, and let $n \geq 0$ and c be fixed integers. Then for any real number $x \in (0, 1]$, we have*

$$
\left| \bar{T}^{(n)}(R, qR + c; x) - \frac{1}{\sqrt{\pi}} \frac{1}{\sqrt[4]{q^2 + \frac{x}{4-x}}} \sqrt{\frac{x}{4-x}} \frac{1}{R^{5/2}} \left(R\sqrt{\frac{x}{4-x}} \right)^n \right.
$$

$$
\left. \times \cos\left[R \arccos\left(1 - \frac{x}{2}\right) - \frac{1}{2}\arctan \frac{1}{q}\sqrt{\frac{x}{4-x}} + \frac{(n-1)\pi}{2} \right] \right| \leq \frac{1}{\sqrt{x}} \bar{M} R^{n-7/2}
$$

(13.15)

$$
\left| \bar{T}'^{(n)}(R, qR + c; x) - \frac{2}{\sqrt{\pi}} \sqrt[4]{q^2 + \frac{x}{4-x}} \frac{1}{R^{1/2}} \left(R\sqrt{\frac{x}{4-x}} \right)^n \right.
$$

$$
\left. \times \cos\left[R \arccos\left(1 - \frac{x}{2}\right) + \frac{1}{2}\arctan \frac{1}{q}\sqrt{\frac{x}{4-x}} + \frac{n\pi}{2} \right] \right| \leq \bar{M}' R^{n-3/2},
$$

(13.16)

for $R \geq \bar{R}_0$, where $\bar{R}_0$, $\bar{M}$ and $\bar{M}'$ are independent of $x \in (0, 1]$.

PROOF. To prove (13.16), substitute (7.13) into (13.13). By Lemma 7.4, all omitted terms in (7.13) give rise to integrals of the type addressed by Proposition 7.2. By the arguments in the proof of Proposition 7.1, the asymptotics of $\bar{T}'^{(n)}(R, qR + c; x)$ comes from the term on the right hand side of (7.13), when $k = n$. More precisely, we obtain

$$
\left| \bar{T}'^{(n)}(R, qR + c; x) - \right.
$$

$$
\left\{ \frac{2q}{\pi} \int_0^1 t^{qR} \frac{t^{c-1/2}}{(1-t)^{1/2}} \left(R\sqrt{\frac{xt}{4-xt}} \right)^n \cos\left[R \arccos\left(1 - \frac{xt}{2}\right) + \frac{n\pi}{2} \right] dt \right.
$$

$$
+ \frac{2}{\pi R} \int_0^1 t^{qR} \frac{t^{c-1/2}}{(1-t)^{1/2}} \left(R\sqrt{\frac{xt}{4-xt}} \right)^{n+1}
$$

$$
\left. \cos\left[R \arccos\left(1 - \frac{xt}{2}\right) + \frac{(n+1)\pi}{2} \right] dt \right\} \Bigg|
$$

$$
\leq \bar{K} R^{n-3/2},
$$

(13.17)

for all $R \geq \bar{\rho}$ and $x \in (0, 1]$, where the constants $\bar{K}$ and $\bar{\rho}$ are independent of $x \in (0, 1]$.

The two integrals on the right hand side of (13.17) can be approximated by means of Proposition 7.1. Carrying this out and using the equation displayed after (7.26) one obtains (13.16). ☐

By Proposition 13.2, the analysis of the asymptotics of $\bar{\omega}_b$ is a repeat of the analysis of the asymptotics of ω_b, with the $T^{(n)}$'s and $T'^{(n)}$'s being replaced by the $\bar{T}^{(n)}$'s and $\bar{T}'^{(n)}$'s, respectively.

Comparing Proposition 13.3 with Proposition 7.1 shows that the only differences between the formulas approximating the $\bar{T}^{(n)}$'s and $\bar{T}'^{(n)}$'s on the one hand and the ones approximating their unbarred versions on the other are:

(i) The numerators 2 and 1 of the first fractions in the approximations are swapped in (13.15)–(13.16) as compared to (7.1)–(7.2);

(ii) The factor $\sqrt{\frac{x}{4-x}}$ in the denominator of the approximant of $T'^{(n)}$ is moved to the denominator of the approximant of the "unprimed" $\bar{T}^{(n)}$;

(iii) The powers of R in the approximants (13.15)–(13.16) are $R^{-5/2}$ and $R^{-1/2}$, as opposed to both being $R^{-3/2}$ in (7.1)–(7.2);

(iv) Decrementation of the argument of the cosine by $\pi/2$ occurs for $\bar{T}^{(n)}$ in (13.15)–(13.16), as opposed to occuring for $T'^{(n)}$ in (7.1)–(7.2).

Find the asymptotics of the moments $\bar{M}_{\alpha_1,\beta_1,\ldots,\alpha_m,\beta_m;\gamma_1,\delta_1,\ldots,\gamma_n,\delta_n}$ by the same reasoning we used for the moments M in Sections 8 and 11. Relations (11.1) and (11.2) change slightly for the present case, to reflect differences (i)–(iv) above. Relation (11.3) stays unchanged. The resulting analog of (11.4) is

$$\bar{M}_{\alpha_1,\beta_1,\ldots,\alpha_m,\beta_m;\gamma_1,\delta_1,\ldots,\gamma_n,\delta_n} = \frac{\bar{E}}{R^{4mn}} \prod_{i=1}^{m}\left(\frac{R_i}{\sqrt{3}}\right)^{\alpha_i+\beta_i} \prod_{j=1}^{n}\left(\frac{R'_j}{\sqrt{3}}\right)^{\gamma_j+\delta_j}$$

$$\times \sum_{\epsilon_1,\ldots,\epsilon_{2m+2n}=\pm1} \frac{1}{D^{\epsilon_1,\ldots,\epsilon_{2m+2n}}} \cos\left\{\frac{R\pi}{3}\left[\sum_{i=1}^{m} A_i(\epsilon_{2i-1}+\epsilon_{2i})+\right.\right.$$

$$\left.+\sum_{j=1}^{n} B_j(\epsilon_{2m+2j-1}+\epsilon_{2m+2j})\right]$$

$$-\sum_{i=1}^{m}\frac{\epsilon_{2i-1}+\epsilon_{2i}}{2}\arctan\frac{1}{q_i\sqrt{3}}+\sum_{j=1}^{n}\frac{\epsilon_{2m+2j-1}+\epsilon_{2m+2j}}{2}\arctan\frac{1}{q'_j\sqrt{3}}$$

$$+\sum_{i=1}^{m}(\epsilon_{2i-1}\alpha_i+\epsilon_{2i}\beta_i)\frac{\pi}{2}+\sum_{j=1}^{n}(\epsilon_{2m+2j-1}\gamma_j+\epsilon_{2m+2j}\delta_j)\frac{\pi}{2}-\sum_{j=1}^{n}(\epsilon_{2j-1}+\epsilon_{2j})\frac{\pi}{2}$$

$$-\sum_{i=1}^{m}\sum_{j=1}^{n}\left(\arctan\frac{\epsilon_{2i-1}A_i+\epsilon_{2m+2j-1}B_j}{q_iA_i+q'_jB_j}+\arctan\frac{\epsilon_{2i-1}A_i+\epsilon_{2m+2j}B_j}{q_iA_i+q'_jB_j}\right.$$

$$\left.\left.+\arctan\frac{\epsilon_{2i}A_i+\epsilon_{2m+2j-1}B_j}{q_iA_i+q'_jB_j}+\arctan\frac{\epsilon_{2i}A_i+\epsilon_{2m+2j}B_j}{q_iA_i+q'_jB_j}\right)\right\}$$

$$+O(R^{-4mn+\sum_{i=1}^{m}(\alpha_i+\beta_i)+\sum_{j=1}^{n}(\gamma_j+\delta_j)-3m-3n-1}),$$

(13.18)

where

$$(13.19) \qquad \bar{E} = \frac{3^m}{2^{2m}\pi^{m+n}} \frac{1}{\prod_{i=1}^{m} R_i^5 \prod_{i=1}^{n} R_i'} \frac{\prod_{j=1}^{n} \sqrt{q_j'^2 + \frac{1}{3}}}{\prod_{i=1}^{m} \sqrt{q_i^2 + \frac{1}{3}}}$$

and $D^{\epsilon_1,\dots,\epsilon_{2m+2n}}$ is given by (11.6).

All consequences of differences (i) and (ii) are reflected in the change of exponents of 2 and 3 in (13.19) versus (11.5). All consequences of difference (iii) are reflected in the changed exponents of R_i and R_i' in (13.19) versus (11.5). The combined effect of differences (i)–(iii) on (11.4) is thus a multiplicative factor independent of $\alpha_1, \beta_1, \dots, \gamma_n, \delta_n$.

The only effect of difference (iv) is that the last sum on the fourth line of (13.18) is $\sum_{j=1}^{n}(\epsilon_{2j-1} + \epsilon_{2j})\frac{\pi}{2}$, as opposed to $\sum_{j=1}^{n}(\epsilon_{2m+2j-1} + \epsilon_{2m+2j})\frac{\pi}{2}$ in (11.4). Furthermore, the statement of Lemma 11.1 is clearly valid also for the multiple sum on the right hand side of (13.18). Therefore, in the multiple sum of (13.18) summation can be restricted to balanced $(\epsilon_1, \dots, \epsilon_{2m+2n})$'s without changing its value. However, this restriction eliminates the only difference between the multiple sums in (13.18) and (11.4). By the arguments that proved (11.22) we obtain

$$\bar{\omega}_b \begin{pmatrix} R_1 & \dots & R_m \,; R_1' & \dots & R_n' \\ v_1 & \dots & v_m \; v_1' & \dots & v_n' \end{pmatrix} = \chi_{2m,2n} \prod_{i=1}^{m} R_i(R_i - 1/2)(R_i + 1/2) \prod_{j=1}^{n} R_j'$$

$$\times \frac{2^{n-m}3^m}{\pi^{m+n}} \frac{1}{\prod_{i=1}^{m} R_i^4 \prod_{j=1}^{n}(R_j')^2} \frac{\prod_{j=1}^{n} \sqrt{(v_j' - c_j')^2 + \frac{1}{3}(R_j')^2}}{\prod_{i=1}^{m} \sqrt{(v_i - c_i)^2 + \frac{1}{3}R_i^2}}$$

$$\times \frac{1}{\prod_{i=1}^{m} \prod_{j=1}^{n} \left[(v_i + v_j' - c_i - c_j')^2 + \frac{1}{3}(R_i - R_j')^2\right]}$$

$$\times \frac{1}{\prod_{i=1}^{m} \prod_{j=1}^{n} \left[(v_i + v_j' - c_i - c_j')^2 + \frac{1}{3}(R_i + R_j')^2\right]}$$

$$\times \frac{(-4)^{m+n}}{3^{m+n}} \prod_{i=1}^{m} R_i^2 \prod_{j=1}^{n}(R_j')^2$$

$$\times \prod_{1 \le i < j \le m} \left[(v_j - v_i)^2 + \frac{1}{3}(R_j - R_i)^2\right] \left[(v_j - v_i)^2 + \frac{1}{3}(R_j + R_i)^2\right]$$

$$\times \prod_{1 \le i < j \le n} \left[(v_j' - v_i')^2 + \frac{1}{3}(R_j' - R_i')^2\right] \left[(v_j' - v_i')^2 + \frac{1}{3}(R_j' + R_i')^2\right]$$

$$(13.20)$$
$$+ O(R^{2m^2 + 2n^2 - 4mn - 2m - 1}).$$

This leads to the following result.

PROPOSITION 13.4. *Let* $R_1, \ldots, R_m, v_1, \ldots, v_m$ *and* $R'_1, \ldots, R'_n, v'_1, \ldots, v'_n$ *depend on* R *as in* (2.3)*. Then as* $R \to \infty$ *the asymptotics of* $\bar{\omega}_b$ *is given by*

$$
\bar{\omega}_b \begin{pmatrix} R_1 & \cdots & R_m & R'_1 & \cdots & R'_n \\ v_1 & & v_m & v'_1 & & v'_n \end{pmatrix} = \bar{\phi}_{2m,2n} \prod_{i=1}^{m}(2R_i) \prod_{j=1}^{n}(2R'_j) \frac{\prod_{j=1}^{n} \sqrt{(R'_j)^2 + 3(v'_j)^2}}{\prod_{i=1}^{m} \sqrt{R_i^2 + 3v_i^2}}
$$

$$
\times \prod_{1 \le i < j \le m} [(R_j - R_i)^2 + 3(v_j - v_i)^2][(R_j + R_i)^2 + 3(v_j - v_i)^2]
$$

$$
\times \frac{\prod_{1 \le i < j \le n}[(R'_j - R'_i)^2 + 3(v'_j - v'_i)^2][(R'_j + R'_i)^2 + 3(v'_j - v'_i)^2]}{\prod_{i=1}^{m} \prod_{j=1}^{n}[(R'_j - R_i)^2 + 3(v'_j + v_i)^2] \,[(R'_j + R_i)^2 + 3(v'_j + v_i)^2]}
$$

(13.21)
$$
+ O(R^{2m^2 + 2n^2 - 4mn - 2m - 1}),
$$

where

(13.22)
$$
\bar{\phi}_{k,l} = \frac{2^l 3^{-(k-l)^2/4 + (3k-l)/4}}{\pi^{(k+l)/2}} \prod_{j=0}^{k-1} \frac{(2)_j}{(1)_j (3/2)_j} \prod_{j=0}^{l-1} \frac{(j+2)_k}{(3/2)_j}.
$$

PROOF. Since by (2.3) the parameters $R_1, \ldots, R_m, v_1, \ldots, v_m$ and $R'_1, \ldots, R'_n$, $v'_1, \ldots, v'_n$ approach infinity as $R \to \infty$, while $c_1, \ldots, c_m$ and $c'_1, \ldots, c'_n$ are constant, it follows that the difference between the product on the right hand side of (13.20) and what it becomes when one omits the additive constants $\pm 1/2$ on its first line and the constants c_i and c'_j is $O(R^{2m^2 + 2n^2 - 4mn - 2m - 1})$. This proves (13.21). Since by assumption the pairs (A_i, q_i), $i = 1, \ldots, m$ of (2.3) are distinct, as well as the pairs (B_j, q'_j), $j = 1, \ldots, n$ of (2.3), the expression on the right hand side of (13.21) has degree $2m^2 + 2n^2 - 4mn - 2m$ in R. Therefore, (13.21) does indeed give the asymptotics of $\bar{\omega}_b$. $\qquad\square$

The proof of Theorem 2.1 now follows readily.

PROOF OF THEOREM 2.1. The statement follows directly from Proposition 3.2, using Theorem 2.2 and Proposition 13.4. $\qquad\square$

A conjectured general two dimensional Superposition Principle

The choice of the side-lengths of the regions (2.1) might seem unmotivated at first, but is in fact quite natural. Indeed, one readily sees that for any lattice hexagon on the triangular lattice, the difference between the lengths of all pairs of opposite sides is the same. Furthermore, this common difference is equal to the difference between the number of up-pointing and down-pointing unit triangles enclosed by the hexagon. Therefore, if we want to choose a lattice hexagon H to contain all our plurimers and possess dimer coverings after their removal, the difference between opposite sides of H has to match the difference $4n + 1 - 4m$ between the number of up-pointing and down-pointing unit triangles in the union of the plurimers. It follows that the sides of H must have the form indicated in Section 2.

By a result of Cohn, Larsen and Propp [8], a dimer covering of a large regular hexagon sampled according to the uniform distribution on all dimer coverings has maximal entropy statistics at its center (and only at its center). This suggests that it is natural to place the hexagon H enclosing the plurimers so that they stay at its center when one scales and lets the size of H grow to infinity. By this observation and the previous paragraph, it follows that from this point of view, up to a translation by an absolute constant (independent of the size of H), (2.1) are the most natural regions to use in the definition (2.2) of the plurimer correlation. (We note that translations by a constant vector that keep the symmetry about ℓ can be handled by exactly the same approach we presented in Sections 2–14.)

In the results of this paper we assumed that all plurimers are triangular. Furthermore, we assumed that all plurimers, except for a single monomer u, have side-length 2, are distributed symmetrically about the say vertical symmetry axis ℓ of u, and that the two plurimer orientations are *separated* by any horizontal h through u, i.e., all plurimers above h point upward, and all below h point downward. However, we conjecture that the Superposition Principle (2.6) is valid for arbitrary plurimers.

To make this conjecture precise, fix a vertex O of the triangular lattice. Consider a rectangular system of coordinates centered at O, with the unit on the horizontal axis equal to the side of a unit triangle, and the unit on the vertical axis equal to twice the height of a unit triangle (these units are chosen so as to maintain consistency with Section 2).

Let $P_1, \ldots, P_n$ be arbitrary plurimers on the triangular lattice (i.e., finite unions of unit triangles), and let $\mathbf{p}_i$ be a distinguished lattice point (a base point) of P_i, $i = 1, \ldots, n$. Write $P_i(R_i, v_i)$ for the translation of the plurimer P_i that takes the base point $\mathbf{p}_i$ to the lattice point of coordinates (R_i, v_i).

As in Section 2, let the charge $\operatorname{ch}(P)$ of the plurimer P be the number of the up-pointing unit triangles of P minus the number of its down-pointing ones. Let H_N be the lattice hexagon centered at O and having side-lengths alternating between N and $N - k$, where $k = \operatorname{ch}(P_1) + \cdots + \operatorname{ch}(P_n)$. Define the correlation $\omega(P_1(R_1, v_1), \ldots, P_n(R_n, v_n))$ by
(14.1)

$$\omega(P_1(R_1, v_1), \ldots, P_n(R_n, v_n)) = \lim_{N \to \infty} \frac{M(H_N \setminus P_1(R_1, v_1) \cup \cdots \cup P_n(R_n, v_n))}{M(H_N \setminus P_1(a_1, b_1) \cup \cdots \cup P_n(a_n, b_n))},$$

where a_i and b_i, $i = 1, \ldots, n$, are some fixed integers specifying a reference position of the plurimers.

We conjecture that the following generalization of Theorem 2.1 holds.

CONJECTURE 14.1. *Suppose the coordinates R_i and v_i, $i = 1, \ldots, n$ are expressed in terms of the integer parameter R as*

$$R_i = A_i R$$
(14.2)
$$v_i = q_i R_i$$

where $0 < A_i \in \mathbb{Q}$, $0 < q_i \in \mathbb{Q}$, $i = 1, \ldots, n$ are all fixed.

Then the asymptotics of the plurimer correlation is given by

$$\omega(P_1(R_1, v_1), \ldots, P_n(R_n, v_n)) = c \prod_{1 \le i < j \le n} [(R_j - R_i)^2 + 3(v_j - v_i)^2]^{\operatorname{ch}(P_i)\operatorname{ch}(P_j)/4}$$

$$+ O(R^{\sum_{1 \le i < j \le n} \operatorname{ch}(P_i)\operatorname{ch}(P_j)/2 - 1})$$

$$= c \prod_{1 \le i < j \le n} \operatorname{d}(P_i(R_i, v_i), P_j(R_j, v_j))^{\operatorname{ch}(P_i)\operatorname{ch}(P_j)/2}$$

(14.3)
$$+ O(R^{\sum_{1 \le i < j \le n} \operatorname{ch}(P_i)\operatorname{ch}(P_j)/2 - 1}),$$

where d is the Euclidean distance, and c depends just on the shapes of the plurimers $P_1, \ldots, P_n$, and not on their coordinates (R_i, v_i).

Writing, for the sake of notational brevity, $P_i = P_i(R_i, v_i)$, it follows from (14.3) that the correlation ω satisfies
(14.4)

$$\omega(P_1, P_2) = c \operatorname{d}(P_1, P_2)^{\operatorname{ch}(P_1)\operatorname{ch}(P_2)/2} + O(R^{\operatorname{ch}(P_1)\operatorname{ch}(P_2)/2 - 1})$$

(14.5)
$$\omega(P_1, \ldots, P_n) = c' \prod_{1 \le i < j \le n} \omega(P_i, P_j) + O(R^{\sum_{1 \le i < j \le n} \operatorname{ch}(P_i)\operatorname{ch}(P_j)/2 - 1}),$$

where c' is a constant independent of R. When taking the logarithm in (14.4), the contribution of the constant is negligible as $R \to \infty$ and we obtain

(14.6)
$$\ln \omega(P_1, P_2) \sim \frac{\operatorname{ch}(P_1)\operatorname{ch}(P_2)}{2} \operatorname{d}(P_1, P_2), \quad R \to \infty.$$

Similarly, provided $\sum_{1 \le i < j \le n} \operatorname{ch}(P_i)\operatorname{ch}(P_j) \ne 0$, when taking the logarithm in (14.5) one can neglect the contribution of the constant and we obtain

(14.7)
$$\ln \omega(P_1, \ldots, P_n) \sim \sum_{1 \le i < j \le n} \ln \omega(P_i, P_j), \quad R \to \infty.$$

Equations (14.6) and (14.7) show that $\ln \omega$ satisfies the characterizing properties of the two-dimensional electrostatic potential—Coulomb's law and the Superposition Principle. Since all classical electrostatics can be deduced from these two properties (see e.g. [**11**, Ch. 4]), Conjecture 14.1 implies that our random tiling model described in Section 2 indeed models classical two-dimensional electrostatics.

We believe that in fact (14.3) holds in still larger generality. There are two ingredients to this extension. First, allow the family of regions used to define the plurimer correlation to be any family with the property that in the scaling limit the plurimers are situated in the region where dimer coverings have maximal entropy statistics. We believe that the correlation defined by means of any such family of regions satisfies (14.3). One instance of this situation is presented in [**5**], where it is shown that (14.3) holds for two charges of magnitudes 2 and -2. An extension found by the author of the result in [**5**] to an arbitrary number of even-side plurimers will be presented in a sequel of the present paper.

Second, based on our result (14.9) below and the conjectured rotational invariance of monomer-monomer correlations (see [**12**]), we conjecture that the natural analog of Conjecture 14.1 on the *square* lattice $\mathbb{Z}^2$ also holds.

More precisely, let $P_1, \ldots, P_n$ be arbitrary plurimers on the square lattice (i.e., finite unions of unit squares). Fix a chessboard coloring of the square lattice, and define the charge $\mathrm{ch}(P)$ of plurimer P to be the difference between the number of its white and black unit squares. As we did for the triangular lattice, consider a base lattice point $\mathbf{p}_i$ in each P_i, and denote by $P_i(x_i, y_i)$ the translation of P_i taking $\mathbf{p}_i$ to (x_i, y_i) (in order for $P_i(x_i, y_i)$ to "look like" P_i, we assume all such translations to be color preserving).

Let AR_N be the "Aztec rectangle region" of sides N and $N + k$ centered at the origin (i.e., the lattice region dual to the corresponding Aztec rectangle graph defined in [**3**]), where $k = \mathrm{ch}(P_1) + \cdots + \mathrm{ch}(P_n)$. Define the correlation of the n plurimers by

$$\omega(P_1(x_1, y_1), \ldots, P_n(x_n, y_n)) = \lim_{N \to \infty} \frac{M(AR_N \setminus P_1(x_1, y_1) \cup \cdots \cup P_n(x_n, y_n))}{M(AR_N \setminus P_1(a_1, b_1) \cup \cdots \cup P_n(a_n, b_n))},$$

where a_i and b_i, $i = 1, \ldots, n$, are some fixed integers specifying a reference position of the plurimers.

CONJECTURE 14.2. *Suppose the coordinates x_i and y_i, $i = 1, \ldots, n$ are expressed in terms of the integer parameter R as*

$$x_i = A_i R$$
$$y_i = q_i x_i,$$

where $0 < A_i \in \mathbb{Q}$, $0 < B_i \in \mathbb{Q}$, $i = 1, \ldots, n$ are all fixed.

Then the asymptotics of the plurimer correlation is given by

$$\omega(P_1(x_1, y_1), \ldots, P_n(x_n, y_n)) = c \prod_{1 \leq i < j \leq n} [(x_j - x_i)^2 + (y_j - y_i)^2]^{\mathrm{ch}(P_i)\,\mathrm{ch}(P_j)/4}$$

$$+ O(R^{\sum_{1 \leq i < j \leq n} \mathrm{ch}(F_i)\,\mathrm{ch}(F_j)/2 - 1})$$

$$= c \prod_{1 \leq i < j \leq n} \mathrm{d}(P_1(x_i, y_i), P_1(x_j, y_j))^{\mathrm{ch}(P_i)\,\mathrm{ch}(P_j)/2}$$

(14.8)
$$+ O(R^{\sum_{1 \leq i < j \leq n} \mathrm{ch}(F_i)\,\mathrm{ch}(F_j)/2 - 1}),$$

where c is a constant depending just on the types of the plurimers $P_1, \ldots, P_n$.

More generally, we believe that any bipartite planar, periodic (i.e., invariant under translations by two non-collinear vectors) graph has an embedding so that joint correlations of plurimers are given by Coulomb's law (using the Euclidean distance) and the Superposition Principle.

For the square lattice, the special case of (14.8) when $n = 2$ and the two plurimers are in fact monomers of opposite color was suggested to hold by Fisher and Stephenson in [**12**]. The further specialization when the second monomer is adjacent to a lattice diagonal through the first was proved by Hartwig [**17**]. A natural extension of Hartwig's result would be to study the correlation of an *arbitrary* collection of monomers along two consecutive lattice diagonals. We state below (see (14.9)) a result we found which is a close analog of this. The proof will appear elsewhere.

For this purpose, rather than phrasing everything in terms of lattice regions and their dimer (domino) coverings, it will be convenient to use the dual set-up of graphs (duals of regions) and their perfect matchings.

Consider therefore the square lattice $\mathbb{Z}^2$ and regard it as a graph. It will be convenient to draw it so that the lattice lines form angles of $45°$ with the horizontal.

We say that a vertex v of the square lattice has been *split* if v is replaced by two new vertices v' and v'' slightly to the left and right of v, respectively, and the four edges incident to v are replaced by two edges joining v' to the former two neighbors of v on its left, and two joining v'' to the former two neighbors of v on its right. Figure 14.1 contains an example of a split vertex.

Color the vertices of the square lattice $\mathbb{Z}^2$ black and white in a chessboard fashion. As above, regard removal of a vertex of $\mathbb{Z}^2$ as creation of a unit charge, of sign determined by its color. In addition, it is natural to consider the operation of splitting a vertex as creating a unit charge *of opposite sign to the one that would be created by removing that vertex* (this is readily seen to be justified if one views perfect matchings as being encoded by families of non-intersecting lattice paths).

Let $m, n \geq 0$ be integers, $\{v_1, \ldots, v_m\}$ and $\{w_1, \ldots, w_n\}$ two disjoint sets of nonnegative integers, and N an even nonnegative integer. We define the graph $AR_N(v_1, \ldots, v_m; w_1, \ldots, w_n)$ to be the subgraph of $\mathbb{Z}^2$ described as follows.

Consider the *Aztec rectangle* (see [**3**]) of width N and height $N + m - n$, i.e., a subgraph of $\mathbb{Z}^2$ consisting of an N by $N + m - n$ array of 4-cycles touching only at vertices. Let ℓ be its vertical symmetry axis, and let O be its vertex on ℓ that is N lattice segments away from its base. Label the vertices on ℓ starting with 0 for O and continuing with consecutive integers as we proceed upward. We define $AR_N(v_1, \ldots, v_m; w_1, \ldots, w_n)$ to be the graph obtained from our Aztec rectangle by removing the vertices labeled v_i, $i = 1, \ldots, m$, and by splitting the vertices w_i, $i = 1, \ldots, n$ (an example is shown in Figure 14.1).

As mentioned above, we regard both kinds of altered vertices as unit charges. We take the reference position of these charges to be when they are packed next to O, and thus define the joint correlation of the charges as

$$\omega(v_1, \ldots, v_m; w_1, \ldots, w_n) :=$$

$$\lim_{N \to \infty} \frac{M(AR_N(v_1, \ldots, v_m; w_1, \ldots, w_n))}{M(AR_N(0, 1, \ldots, m-1; m, m+1, \ldots, m+n-1))}.$$

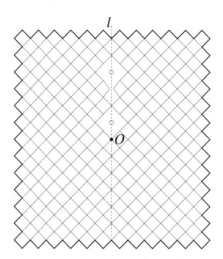

Figure 14.1. $AR_{12}(1, 4; 3)$.

Using results from [3] and doing the asymptotic analysis of the expressions they lead to, we found that
(14.9)
$$\omega(v_1, \ldots, v_m; w_1, \ldots, w_n) \sim c_{m,n} \frac{\prod_{1 \le i < j \le m}(v_j - v_i)^{1/2} \prod_{1 \le i < j \le n}(w_j - w_i)^{1/2}}{\prod_{i=1}^{m} \prod_{i=1}^{n} |v_i - w_j|^{1/2}},$$

as the distances between charges grow to infinity keeping constant mutual ratios, where $c_{m,n}$ is some explicit constant depending only on m and n. The above formula shows that the Superposition Principle holds on the square lattice for *any* distribution of unit charges along a lattice diagonal.

As a particular case of (14.9), we obtain, after completing the fairly laborious task of working out the constant, that

(14.10)
$$\omega(0, d) \sim \frac{\pi \sqrt{e}}{2^{1/3} A^6} \sqrt{d}, \quad d \to \infty$$

where $A = 1.282427\ldots$ is the Gleisher-Kinkelin constant [14],

$$1^1 2^2 \cdots n^n \sim A n^{n^2/2 + n/2 + 1/12} e^{-n^2/4}.$$

Formula (14.10) is a counterpart of the result of Hartwig [17], which addresses the case of vertices of opposite colors removed from adjacent diagonals.

REMARK 14.2. After scaling, the above set-up places the momomers whose correlation is measured exactly in the center of the scaled Aztec rectangle, which in the limit approaches an Aztec diamond. According to a result of Cohn, Elkies and Propp [7], a uniformly sampled tiling of an Aztec diamond has maximal entropy statistics at its center, and only at its center. The above set-up seems therefore natural.

Three dimensions and concluding remarks

As far as the author knows, the question of working out the joint correlations of an arbitrary number of holes for the dimer model on a bipartite lattice—and thereby establishing the emergence of electrostatics in this way—was not considered before in the literature. Perhaps the case of removing several regions from a lattice and study their joint correlations was not considered before due to the complexity the problem presents already in the case of two removed monomers (see [12]). This might have been also the reason for not considering the case of two removed regions of arbitrary charges, a situation that already points to the phenomena of electrostatics. Another possible reason is that the analysis of [12] was done on the square lattice, while the hexagonal lattice is the one on which natural regions to remove—triangular regions—can have arbitrary charges. (Indeed, on the square lattice to get a region of charge s one needs for instance to line up s diagonally adjacent monomers. Removing a whole dimer is natural on the square lattice, and this was studied extensively in [12]. However, since dimers have charge zero, the study of their correlations does not detect the Superposition Principle!)

We mention that there are results in the physics literature outlining some connections of the Ising model, a well studied statistical physical model, and electrostatics (e.g. Hurst and Green note in [18] that the relations satisfied by certain matrices they employ to solve the Ising model "are similar to the relations satisfied by fermion emission and absorption matrices," and in [30], which presents exact calculations of 2-point spin-spin correlations in the Ising model, Wu, McCoy, Tracy and Barouch are led to certain integral equations that arose before in work of Myers [26] in the context of electromagnetic scattering from a strip; in [25] McCoy, Perk and Wu give recurrences for n-point spin correlations, but without analyzing their asymptotics). Since the Ising model on a lattice can be equivalently phrased as a dimer model on a modified, suitably weighted lattice (see e.g. [19]), this yields a connection between electrostatics and the dimer model. However, this turns out to lead to a totally different object of study in the dimer model than the joint correlation of holes we considered in this paper: two spins being correlated in an Ising state turns out to correspond to requiring two faces in the corresponding dimer lattice to be on the same side of the union of certain polygons corresponding to the Ising state. Moreover, the fundamental difference (4) mentioned in the Introduction between our model and the ones surveyed in [27] holds also between our model and the Ising model.

One key feature of the work described above is that even though the quantities that need to be studied asymptotically are not "round" (i.e., do not factor as products of small factors) in general, they can be expressed as multiple sums of round terms, courtesy of the exact enumeration formulas from Part B of this Memoir and [3]. These multiple sums can then, in certain situations, be reduced to single or double sums that in turn lead to special functions whose asymptotics can be obtained using specific techniques such as Laplace's method. This view supplies additional motivation for the already well-represented study of regions whose tilings are enumerated by simple product formulas.

The question of studying joint correlations of missing vertices in lattice graphs can naturally be phrased also in dimensions other than two. The really compelling case is that of three dimensions. *Does there exist a three dimensional analog of our model that would model classical electrostatics in the physical, three dimensional space?*

We present such a possibility in Question 15.1 below. In the case of an affirmative answer to this question, we show in Remark 15.2 how a natural new parameter could be introduced in our model so that the parallel between the three dimensional analogs of (2.13) and (2.14) holds for any temperature T.

Based on the two dimensional case, we are guided in the phrasing of (15.2) by the assumption that in three dimensions as well an analogous random tiling model would manifest the phenomena of electrostatics.

The square lattice version of our model is easiest to parallel in three dimensions. Consider $\mathbb{R}^3$ divided into unit cubes by the lattice $\mathbb{Z}^3$. Color the unit cubes black and white so that adjacent cubes have different colors. Regard the unit cubes as monomers, and their finite unions as plurimers.

Consider two monomers $a = (0,0,0)$ and $b_r = (0,0,2r-1)$ of opposite color and include them in a large cube C_N, where N is even. Based on our guiding physical intuition, we wish to define their correlation $\omega(a, b_r)$ to be proportional to $\mathrm{M}(C_N \setminus \{a, b_r\})$, for large N, in such a way that $-\ln \omega(a, b_r)$ behaves asymptotically like the potential energy of two unit charges of opposite sign at distance $2r$, i.e., like a positive constant times $-1/(2r)$. In particular, we should have $\omega(a, b_r) \to 1$, as $r \to \infty$.

This suggests that in the definition of $\omega(a, b_r)$ we should normalize by letting the monomers be far apart. Based on this we define plurimer correlation as follows.

Fix a lattice point O, and consider a rectangular system of coordinates centered at O. Let P_i be an arbitrary plurimer, and let $\mathbf{p}_i \in \mathbb{Z}^3$ be a base point of P_i, for $i = 1, \ldots, n$. For any integer $R \geq 1$, define RP_i to be the translation of P_i that takes $\mathbf{p}_i$ to $R\mathbf{p}_i$.

Enclose the n plurimers in a large cube C_N of side N, centered at O. For the sake of definition simplicity, assume that $\mathrm{ch}(P_1) + \cdots + \mathrm{ch}(P_n) = 0$ (this condition is not essential for defining the correlation below and phrasing Question 15.1—it can be circumvented by replacing the enclosing cubes C_N by a suitable family of regions D_N with the property that $\mathrm{ch}(D_N) = \mathrm{ch}(P_1) + \cdots + \mathrm{ch}(P_n)$; for $\mathbb{Z}^2$, this was accomplished in Section 14 by choosing Aztec rectangles AR_N).

Assuming the following limit exists, define the correlation $\omega(P_1, \ldots, P_n)$ of the plurimers by

$$(15.1) \qquad \omega(P_1, \ldots, P_n) = \lim_{R \to \infty} \lim_{N \to \infty} \frac{\mathrm{M}(C_N \setminus P_1 \cup \cdots \cup P_n)}{\mathrm{M}(C_N \setminus RP_1 \cup \cdots \cup RP_n)},$$

where $M(D)$ denotes the number of dimer coverings of the lattice region (on $\mathbb{Z}^3$) D. (The inside limit should exist for any R—it is just the ratio of two joint correlations of plurimers in a sea of dimers; we are assuming in addition that its value approaches a limit as $R \to \infty$.)

The possibility we referred to above for modeling three dimensional electrostatics is phrased below. As in two dimensions, the charge $\mathrm{ch}(P)$ of a plurimer P is defined to be the difference between the numbers of its white and black unit cubes.

QUESTION 15.1. *Is it true that*

$$(15.2) \qquad \omega(RP_1,\dots,RP_n) = 1 - k_3 \sum_{1\le i<j\le n} \frac{\mathrm{ch}(P_i)\,\mathrm{ch}(P_j)}{\mathrm{d}(R\mathbf{p}_i, R\mathbf{p}_j)} + O(R^{-2})$$

as $R \to \infty$, where $k_3 > 0$ is an absolute constant?

If the answer to this question is affirmative, taking the logarithm in (15.2) we obtain

$$(15.3) \qquad \ln \omega(RP_1,\dots,RP_n) \sim -k_3 \sum_{1\le i<j\le n} \frac{\mathrm{ch}(P_i)\,\mathrm{ch}(P_j)}{\mathrm{d}(R\mathbf{p}_i, R\mathbf{p}_j)}, \qquad R \to \infty.$$

This would show then that indeed $\ln \omega$ behaves like the electrostatic potential in three dimensions.

REMARK 15.2. Assuming the answer to Question 15.1 is affirmative, we can introduce a new parameter in our model that allows it to parallel electrostatics for any temperature.

The new parameter, denoted by x, ranges over the odd positive integers. For any such x, refine each unit cube c of $\mathbb{Z}^3$ into x^3 equal smaller cubes, and properly color the latter black and white so that the smaller cubes fitting in the corners of c have the same color as c (this can clearly be done, since x is odd). Moreover, if we regard the subdivided cube c as a plurimer on the lattice $(\frac{1}{x}\mathbb{Z})^3$, its charge is readily seen to agree with the charge of c, regarded as a monomer on $\mathbb{Z}^3$ (again, this is due to x being odd).

Let P_i and $\mathbf{p}_i$, $i = 1,\dots,n$, be as above. Define the correlation $\omega_x(P_1,\dots,P_n)$ by

$$(15.4) \qquad \omega_x(P_1,\dots,P_n) = \lim_{R\to\infty}\lim_{N\to\infty} \frac{\mathrm{M}_x(C_N \setminus P_1 \cup \dots \cup P_n)}{\mathrm{M}_x(C_N \setminus RP_1 \cup \dots \cup RP_n)},$$

where for a lattice region H in $\mathbb{Z}^3$, $M_x(H)$ denotes the number of its dimer coverings when regarded as a lattice region in $(\frac{1}{x}\mathbb{Z})^3$. (Since as noted in the previous paragraph our lattice refinement preserves charge, H has the same number of black and white fundamental regions when considered in $(\frac{1}{x}\mathbb{Z})^3$, provided it does so when considered in $\mathbb{Z}^3$.) Clearly, existence of the limit (15.1) implies existence of the above limit.

The asymptotics of $\omega_x(P_1,\dots,P_n)$ can be deduced from (15.1) as follows. Scale down the lattice $\mathbb{Z}^3$ by a linear factor of x, keeping our lattice point O fixed, and view the plurimers and their base points as being in the scaled down lattice: the plurimers become P_i^x, $i = 1,\dots,n$, and their base points $\mathbf{p}_i^x = x\mathbf{p}_i$, $i = 1,\dots,n$. It follows from the definition that

$$\omega_x(P_1,\dots,P_n) = \omega(P_1^x,\dots,P_n^x).$$

Clearly, we also have $\mathrm{d}(R\mathbf{p}_i^x, R\mathbf{p}_j^x) = x\,\mathrm{d}(R\mathbf{p}_i, R\mathbf{p}_j)$, for all $1 \le i < j \le n$. Furthermore, the plurimer $(RP_i)^x$ is the same as the plurimer RP_i^x. Therefore, we obtain by (15.2) that

$$(15.5) \qquad \omega_x(RP_1, \ldots, RP_n) = 1 - \frac{k_3}{x} \sum_{1 \le i < j \le n} \frac{\mathrm{ch}(P_i)\,\mathrm{ch}(P_j)}{\mathrm{d}(R\mathbf{p}_i, R\mathbf{p}_j)} + O(R^{-2}).$$

Taking the logarithm in both sides we deduce from (15.5) that

$$(15.6) \qquad \ln \omega_x(RP_1, \ldots, RP_n) \sim -\frac{k_3}{x} \sum_{1 \le i < j \le n} \frac{\mathrm{ch}(P_i)\,\mathrm{ch}(P_j)}{\mathrm{d}(R\mathbf{p}_i, R\mathbf{p}_j)}, \qquad R \to \infty.$$

Repeating the analysis at the end of Section 2, (15.6) implies that in the limit $N \to \infty$, when one samples uniformly at random from all dimer coverings of the x-subdivided $C_N \setminus P_1 \cup \cdots \cup P_n$, the relative probability of having the plurimers at preassigned distances d_{ij} versus d'_{ij} is, by the fundamental theorem of statistical mechanics (see e.g. [10, §40)]), *exactly the same* as the relative probability of having n corresponding electrical charges at those distances, at temperature

$$(15.7) \qquad T = xq_e^2/(4\pi\epsilon_0 kk_3 L_P),$$

where q_e is the charge of the electron, k is Boltzmann's constant, L_P is the Planck length ($\sim 10^{-35}$, smallest non-zero length that "makes sense" physically; in our analysis above, we express all physical distances using L_P as a unit[10]) and ϵ_0 is the permittivity of empty space.

The free energy per unit volume in this model equals the free energy per $\mathbb{Z}^3$-site in the dimer model on the lattice $(\frac{1}{x}\mathbb{Z})^3$. This energy is $F = l_3 x^3$, where $l_3 \sim 0.446\ldots$ is the three-dimensional dimer constant (see e.g. [16][4]).

As mentioned in the footnote at the end of Section 2, there is an extra "calibration" parameter one can consider in our model—a fixed positive integer a so that a physical elementary charge corresponds to a plurimer of charge a in our model. The effect of this extra parameter in (15.7) is to multiply its right hand side by $1/a^2$. Expressing x from the resulting relation and substituting into $F = l_3 x^3$ we obtain

$$(15.8) \qquad F = \left(\frac{4\pi\epsilon_0 kk_3 L_P a^2}{q_e^2}\right)^3 l_3 T^3.$$

This is similar in form to the formula for the free energy per unit volume, for high temperature T, obtained in quantum field theory. Note that the constant of proportionality in (15.8) is $\sim (10^{-30} a^2)^3$, so the value of F depends crucially on a; for small a the value is negligible even for very large temperatures T, while for $a \sim 10^{15}$—a conceivable value, given that $L_P \sim 10^{-35}$—it becomes quite significant. The free energy per unit volume F in our model is reminiscent of the cosmological constant, which also encodes in some sense the overall energy stored in the vacuum of space. The counterpart in our model of the fact that the cosmological constant is believed to be positive but too small to be detected by current experiments would then be that the calibration parameter a satisfies $a \ll 10^{15}$.

It is important to remark that while the Superposition Principle seems to hold independently of the background bipartite lattice, the effect of lattice refinement

---

[10]Strictly speaking, one needs to express physical distances in terms of L_P also in (2.13), but since (2.13) is invariant under changing the unit for distance, this was not necessary in the two dimensional case.

on both (2.13) and its three dimensional analog that (15.2) would imply depends crucially on the lattice. For instance, (14.8) would imply that (2.13) holds also on the square lattice. However, in this case x-fold refinement does not provide an extra parameter: for even x the plurimers become neutral on the refined lattice, while for odd x the plurimer charges are invariant under refinement, a fact which together with the invariance of (2.13) under distance scaling shows that (2.13) does not change in this case under refinement. Similarly, for bipartite lattices in three dimensions the charge is not invariant in general under refinement, so different lattices lead to different analogs of (15.7) and (15.8).

An interesting question is the following: if bipartite lattices lead to electrostatic effects, what do non-bipartite lattices lead to? Since the latter do not have a black and white coloring, it is natural to expect uniform behavior—either universal attraction, or universal repelling. Is one perhaps led to the effects of some other fundamental physical force? The most natural non-bipartite plane lattice, the triangular lattice, is considered in [9], and an analysis of the monomer-monomer correlations on it, paralleling that of Fisher and Stephenson [12] on the square lattice, is carried out. The first 15 correlations of two monomers in adjacent lattice lines are computed, and their values [9, Table III] provide convincing evidence for an exponential decay to a limiting positive constant. Changing the normalization in the definition of correlation in [9, p.8] so that one normalizes by the reference position of two infinitely separated monomers (as in (15.4)), this is equivalent to the modified correlation exponentially approaching zero. Then the "potential" would be $\ln \omega \sim -kr$, with $k > 0$ a constant and r the separation between monomers, and the "field" $\frac{d}{dr} \ln \omega$—or more directly, according to "discrete calculus" (see e.g. [15]), $\omega(r+1)/\omega(r) - 1$—would approach $-k$. This is similar to the case of interquark force, which approaches a constant as the separation between the quarks approaches infinity. It would be very interesting to study the joint correlation of several plurimers on the triangular lattice in more detail, in particular to determine whether it is a function depending just on the pair correlations (as it is the case in the presence of a Superposition Principle), and to find out how it depends on the plurimer sizes.

We conclude by mentioning that since the set of all tilings of a region is equivalent to the set of all families of non-intersecting lattice paths with certain starting and ending points—which is in turn equinumerous, by the Gessel-Viennot theorem, with the set of all (appropriately signed) families of lattice paths with these starting and ending points— averaging over all such tilings is reminiscent of the "sum-over-paths" interpretation of particles in quantum mechanics due to Feynman. The results in this paper show that if the different tilings of a region with holes are associated with different ways for pairs of virtual particles and their antimatter companions to annihilate, then the microscopic frenzy of quantum-mechanical fluctuations in the vacuum of empty space that Feynman once jokingly described as "Created and annihilated, created and annihilated—what a waste of time," is seen to actually generate the phenomena of electrostatics[11].

---

[11] A consequence of quantum mechanics is that the quantum fluctuations of the vacuum drive the intrinsic strength of the electric field of charged particles to get larger when examined on short distance scales. This offers the opportunity to perform a test for the parallel between our model and electrostatics. Explicit numerical calculations based on results of this paper and [5] confirm that the quantity $\omega(r+1)/\omega(r) - 1$ (where $\omega(r)$ is the correlation between a fixed hole and another hole of fixed shape at distance r), the analog in our model of the electric field, when divided by the

Acknowledgments. I thank Richard Stanley for his continuing interest in this work, his encouragements and many helpful discussions. I thank Timothy Chow for many interesting conversations and my brother Alexandru Ciucu for following enthusiastically the development of the ideas presented in this paper. I also thank the anonymous referee for the thorough review and for helpful suggestions.

---

two dimensional electric field intensity $1/r$, does indeed yield a quantity that grows as r decreases to 0. Details will appear in a sequel to this paper.

Bibliography

[1] L. V. Ahlfors, *Complex analysis; an introduction to the theory of analytic functions of one complex variable*, McGraw-Hill, New York, 1966.

[2] P. G. Carter, *An empirical equation for the resonance energy of polycyclic aromatic hydrocarbons*, Trans. Faraday. Soc. **45** (1949), 597–602.

[3] M. Ciucu, *Enumeration of perfect matchings in graphs with reflective symmetry*, J. Comb. Theory Ser. A **77** (1997), 67–97.

[4] M. Ciucu, *An improved upper bound for the three dimensional dimer problem*, Duke Math. J. **94** (1998), 1–11.

[5] M. Ciucu, *Rotational invariance of quadromer correlations on the hexagonal lattice*, Adv. in Math. **191** (2005), 46–77.

[6] M. Ciucu, *The scaling limit of the correlation of holes on the triangular lattice under periodic boundary condition*, http://arxiv.org/abs/math-ph/0501071, 84p.

[7] H. Cohn, N. Elkies and J. Propp, *Local statistics for random domino tilings of the Aztec diamond*, Duke Math. J. **85** (1996), 117–166.

[8] H. Cohn, M. Larsen and J. Propp, *The shape of a typical boxed plane partition*, New York J. of Math. **4** (1998), 137–165.

[9] P. Fendley, R. Moessner and S. L. Sondhi, *The classical dimer model on the triangular lattice*, Phys. Rev. B **66** (2002), 214513.

[10] R. P. Feynman, *The Feynman Lectures on Physics, vol. I*, Addison-Wesley, Reading, Massachusetts, 1963.

[11] R. P. Feynman, *The Feynman Lectures on Physics, vol. II*, Addison-Wesley, Reading, Massachusetts, 1964.

[12] M. E. Fisher and J. Stephenson, *Statistical mechanics of dimers on a plane lattice. II. Dimer correlations and monomers*, Phys. Rev. (2) **132** (1963), 1411–1431.

[13] I. M. Gessel and X. Viennot, *Binomial determinants, paths, and hook length formulae*, Adv. in Math. **58** (1985), 300–321.

[14] I. S. Gradshtein and I. M. Ryzhik, *Table of integrals, series, and products*, Academic Press, New York, 1980.

[15] R. L. Graham, D. E. Knuth and O. Patashnik, *Concrete Mathematics*, Addison-Wesley, Reading, Mass., 1989.

[16] J. M. Hammersley, *An improved lower bound for the multidimensional dimer problem*, Proc. Cambridge Philos. Soc. **64** (1968), 455–463.

[17] R. E. Hartwig, *Monomer pair correlations*, J. Mathematical Phys. **7** (1966), 286–299.

[18] C. A. Hurst and H. S. Green, *New solution of the Ising problem for a rectangular lattice*, J. Chem. Phys. **33** (1960), 1059–1062.

[19] P. W. Kasteleyn, *Dimer statistics and phase transitions*, J. Math. Phys. **4** (1963), 287–293.

[20] R. Kenyon, *Local statistics of lattice dimers*, Ann. Inst. H. Poincaré Probab. Statist. **33** (1997), 591–618.

[21] R. Kenyon, *Long-range properties of spanning trees. Probabilistic techniques in equilibrium and nonequilibrium statistical physics*, J. Math. Phys. **41** (2000), 1338–1363.

[22] Werner Krauth and R. Moessner, *Pocket Monte Carlo algorithm for classical doped dimer models*, Phys. Rev. B **67** (2003), 064503.

[23] L. Lovász and M. D. Plummer, *Matching theory*, Elsevier Science Publishers, Amsterdam, 1986.

[24] Y. L. Luke, *The special functions and their approximations*, Academic Press, New York, 1969.

[25] B. M. McCoy, J. H. H. Perk and T. T. Wu, *Ising field theory: Quadratic difference equations for the n-point Green's functions on the lattice*, Phys. Rev. Lett. **46** (1981), 757–760.

[26] J. Myers, *Wave scattering and the geometry of a strip*, J. Math. Phys. **6** (1965), 1839–1846.

[27] B. Nienhuis, *Coulomb gas formulation of two-dimensional phase transitions*, Phase Transitions **11** (1987), Academic Press, London, 1–53.

[28] F. W. J. Olver, *Asymptotics and special functions*, Academic Press, New York, 1974.

[29] J. R. Stembridge, *Nonintersecting paths, Pfaffians and plane partitions*, Adv. in Math. **83** (1990), 96–131.

[30] T. T. Wu, B. M. McCoy, C. A. Tracy and E. Barouch, *Spin-spin correlation functions for the two-dimensional Ising model: Exact theory in the scaling region*, Phys. Rev. B **13** (1976), 316–374.

Part B

Plane partitions I: a generalization of MacMahon's formula

Introduction

A plane partition is a rectangular array of nonnegative integers with the property that all rows and columns are weakly decreasing. A plane partition $\pi = (\pi_{ij})_{0 \leq i < a, 0 \leq j < b}$ can be identified with its three dimensional diagram

$$D_\pi = \{(i, j, k) : 0 \leq k < \pi_{ij}\},$$

and hence can be viewed as an order ideal of $\mathbb{N}^3$ (an order ideal of a partially ordered set is a subset I such that $x \in I$ and $y \leq x$ imply $y \in I$).

Replacing each point of D_π by a unit cube centered at it and parallel to the coordinate axes, the diagram of π becomes a stack of unit cubes justified onto the three coordinate planes. By projection on a plane normal to the vector $(1, 1, 1)$ one obtains that plane partitions whose diagrams fit inside an $a \times b \times c$ box are identified with tilings of a hexagon of side-lengths a, b, c, a, b, c (in cyclic order) and angles of 120 degrees by unit rhombi with angles of 60 and 120 degrees (see [3] or [7] for the details of this bijection); we call such rhombi *lozenges* and such tilings *lozenge tilings*.

MacMahon showed [8] that the number of plane partitions contained in an $a \times b \times c$ box (equivalently, the number of lozenge tilings of the corresponding hexagon), with $a \leq b$, equals

$$\frac{(c+1)(c+2)^2 \cdots (c+a)^a (c+a+1)^a \cdots (c+b)^a (c+b+1)^{a-1} \cdots (c+b+a-1)}{1 \cdot 2^2 \cdots a^a \cdot (a+1)^a \cdots b^a \cdot (b+1)^{a-1} \cdots (b+a-1)}.$$

The elegance of this result strongly invites one to find generalizations of it. Two natural directions in which attempts have been made are the following. First, one may consider enumerating order ideals of the product of d finite chains, with $d \geq 4$ (the case $d = 4$ was already considered by MacMahon [9]). Numerical evidence suggests that there is no simple product formula for the answer to this problem, and very little is known about it (see [11, I §11]). Second, taking the viewpoint of tilings, one may consider the question of counting tilings by unit rhombi of $2d$-gons with parallel and congruent opposite sides. Again, there seem to be very few results on this in the literature for $d \geq 4$ (see [4]).

We generalize MacMahon's result in the case $b = c$ by presenting a general family of hexagonal regions with holes on the triangular lattice, such that the number of tilings of each member of this family is given by a simple product formula (examples of regions from this family are shown in Figures 1.2–1.4).

Consider the tiling of the plane by unit equilateral triangles. Define a *region* to be any region of the plane that can be obtained as a finite union of these unit triangles.

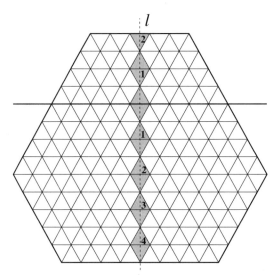

FIGURE 1.1. The hexagon $H(5,5,3)$; its vertebrae are shaded
and labeled starting from the indicated horizontal reference line.

Consider the hexagonal region $H(a,b,k)$ with side-lengths $a, b+k, b, a+k, b, b+k$, where k is a nonnegative integer, and suppose $H(a,b,k)$ is drawn such that its base is the edge of length $a+k$ (see Figure 1.1 for an example). Then $H(a,b,k)$ has a vertical symmetry axis ℓ.

It is easy to see that $H(a,b,k)$ contains precisely k more unit triangles pointing upward than unit triangles pointing downward. Since a lozenge always covers one unit triangle of each kind, it follows that $H(a,b,k)$ does not have any tilings for $k \neq 0$.

This can be fixed by removing some unit triangles from $H(a,b,k)$, so as to restore the balance between up- and down-pointing unit triangles. One natural way to do this, which was in fact our motivating example, is to remove an up-pointing triangular region of side-length k from the center of $H(a,b,k)$ (the instance $k = 1$ of this is treated in [**2**] and corresponds to a problem posed independently by Propp [**10**] and Kuperberg (private communication)). At least from the viewpoint of our proof, it turns out that it is easier to enumerate tilings of the more general regions defined below.

We call a triangular subregion of $H(a,b,k)$ which is symmetric with respect to ℓ a *window*. If the vertex opposite the base of a window is above the base we call the window a Δ-*window*; otherwise we call it a ∇-*window*. A window is called even or odd according as its side-length is even or odd.

The unit triangles of $H(a,b,k)$ along ℓ can be grouped in pairs forming lozenges — we call them *rhombic vertebrae* — except possibly for those such triangles touching the boundary of $H(a,b,k)$, which we call *triangular vertebrae* (instances of both kinds of vertebrae occur in Figure 1.1; the vertebrae are indicated by a shading). The definitions below involve the concept of labeling the vertebrae starting from a reference line — a line of the triangular grid perpendicular to ℓ: by this we mean that we label the vertebrae that fully lie on each side of the reference line starting with 1 for the closest vertebra, 2 for the next, and so on; see Figure 1.1 (in case the

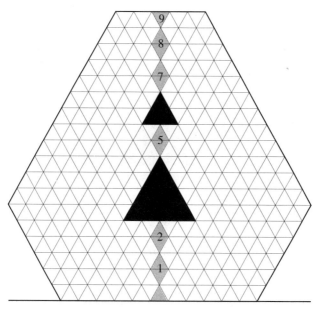

FIGURE 1.2. The region $H_{(1,2,5,7,8,9)}(5,6,6)$. The vertebrae contained in it are shaded, and their labels and the reference line for the labeling are shown.

reference line is the base of $H(a,b,k)$ and the vertebra touching it is triangular, we skip this vertebra and start by labeling the next one by 1).

Suppose k is even. Remove from $H(a,b,k)$ some number of even Δ-windows of total size k (the total size of a collection of windows is the sum of the side-lengths of the windows). Since we removed only even windows, each vertebra is either fully removed or fully left in place. Label the vertebrae of $H(a,b,k)$ starting from its base, and suppose the vertebrae in the leftover region have, in increasing order, labels $l_1, \ldots, l_n$ (see Figure 1.2 for an example). Let $\mathbf{l} = (l_1, \ldots, l_n)$. The region obtained from $H(a,b,k)$ by removing windows as above may have forced lozenges (e.g., if some of the windows touch). Remove all such forced lozenges. Denote the leftover region by $H_{\mathbf{l}}(a,b,k)$.

If k is odd, start, from top to bottom, by removing a number of even Δ-windows from $H(a,b,k)$, then remove a single odd window (which can be either a Δ- or a ∇-window) and continue downwards by removing some even ∇-windows (the set of removed even windows above the odd window, or the set of removed windows below it, or possibly both, may be empty). In this process, obey the restriction that the total size of the Δ-windows is k larger than the total size of the ∇-windows. Choose the base of the single odd window as our reference line, and label the vertebrae on both sides of it as indicated above. Suppose the vertebrae below the reference line in the leftover region have, in increasing order, labels $l_1, \ldots, l_m$, and the ones above it have, in increasing order, labels $q_1, \ldots, q_n$ (note that one or both of these lists may be empty). Let $\mathbf{l} = (l_1, \ldots, l_m)$, $\mathbf{q} = (q_1, \ldots, q_n)$. Remove any forced lozenges, and denote the leftover region by $H_{\mathbf{l},\mathbf{q}}(a,b,k)$ or $\bar{H}_{\mathbf{l},\mathbf{q}}(a,b,k)$, according as the removed odd window was a Δ- or a ∇-window (Figures 1.3 and 1.4 show an example of each kind).

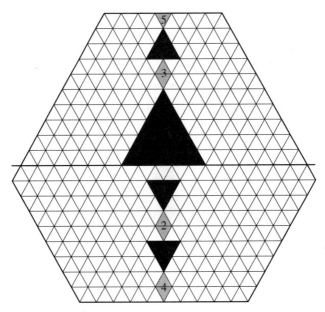

FIGURE 1.3. The region $H_{(2,4),(3,5)}(7,8,3)$. The vertebrae
contained in it, their labels and the reference line are shown.

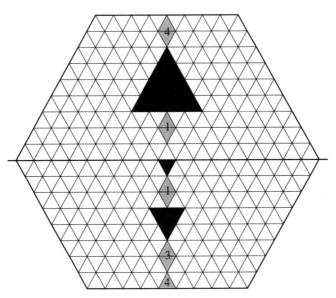

FIGURE 1.4. The region $\bar{H}_{(1,3,4),(1,4)}(8,8,1)$,
with the vertebrae it contains.

The regions $H_{\mathbf{l}}(a,b,k)$, $H_{\mathbf{l},\mathbf{q}}(a,b,k)$, and $\bar{H}_{\mathbf{l},\mathbf{q}}(a,b,k)$ are the regions whose
number of tilings we will determine, thus generalizing the case $b = c$ of MacMahon's
result (which is obtained for $k = 0$).

Before giving the formulas expressing the number of tilings of these regions, we need to define two families of polynomials, both indexed by pairs of lists of positive integers.

For $m, n \geq 0$ define monic polynomials $B_{m,n}(x)$ and $\bar{B}_{m,n}(x)$ by the formulas

$$B_{m,n}(x) = 2^{-mn-m(m-1)/2}(x+n+1)_m(x+n+2)_m$$

$$\times (x+2)(x+3)^2 \cdots (x+n-1)^2(x+n)$$

$$\times \left(x+\frac{3}{2}\right)\left(x+\frac{5}{2}\right)^2 \cdots \left(x+\frac{2n-1}{2}\right)^2\left(x+\frac{2n+1}{2}\right)$$

$$(1.1) \qquad \times \prod_{i=1}^{n}\frac{(x+i)_m}{(x+i+1/2)_m}\prod_{i=1}^{m}(2x+n+i+2)_{n+i-1}$$

$$\bar{B}_{m,n}(x) = 2^{-mn-n(n+1)/2}(x+m+1)_n$$

$$\times (x+1)(x+2)^2 \cdots (x+m-1)^2(x+m)$$

$$\times \left(x+\frac{3}{2}\right)\left(x+\frac{5}{2}\right)^2 \cdots \left(x+\frac{2m-3}{2}\right)^2\left(x+\frac{2m-1}{2}\right)$$

$$(1.2) \qquad \times \prod_{i=1}^{m}\frac{(x+i)_n}{(x+i+1/2)_n}\prod_{i=1}^{n}(2x+m+i+1)_{m+i},$$

where $(a)_k := a(a+1)\ldots(a+k-1)$ is the shifted factorial (in the middle two lines of (1.1) and (1.2), the bases grow by one from each factor to the next, while the exponents increase by one midway through and then decrease by one to the end).

Next, given (possibly empty) lists of strictly increasing positive integers $\mathbf{l} = (l_1,\ldots,l_m)$ and $\mathbf{q} = (q_1,\ldots,q_n)$, define constants $c_{\mathbf{l},\mathbf{q}}$ and $\bar{c}_{\mathbf{l},\mathbf{q}}$ by setting

(1.3)
$$c_{\mathbf{l},\mathbf{q}} = 2^{\binom{n-m}{2}-m}\prod_{i=1}^{m}\frac{1}{(2l_i)!}\prod_{i=1}^{n}\frac{1}{(2q_i-1)!}\frac{\prod_{1\leq i<j\leq m}(l_j-l_i)\prod_{1\leq i<j\leq n}(q_j-q_i)}{\prod_{i=1}^{m}\prod_{j=1}^{n}(l_i+q_j)}$$

(1.4)
$$\bar{c}_{\mathbf{l},\mathbf{q}} = 2^{\binom{n-m}{2}-m}\prod_{i=1}^{m}\frac{1}{(2l_i-1)!}\prod_{i=1}^{n}\frac{1}{(2q_i)!}\frac{\prod_{1\leq i<j\leq m}(l_j-l_i)\prod_{1\leq i<j\leq n}(q_j-q_i)}{\prod_{i=1}^{m}\prod_{j=1}^{n}(l_i+q_j)}.$$

Let $\lambda = \lambda(\mathbf{l})$ and $\mu = \mu(\mathbf{q})$ be the partitions having parts $l_m - m, l_{m-1} - m + 1,\ldots,l_1 - 1$ and $q_n - n, q_{n-1} - n + 1,\ldots,q_1 - 1$, respectively.

We are now ready to define our polynomials $P_{\mathbf{l},\mathbf{q}}(x)$ and $\bar{P}_{\mathbf{l},\mathbf{q}}(x)$. They are given by the following formulas:

$$P_{1,\mathbf{q}}(x - \lambda_1) = c_{1,\mathbf{q}} B_{m,n}(x)$$
(1.5)
$$\times \prod_{c \in \lambda}(x - h(c) + m + 1)(x + h(c) + n + 1) \prod_{c \in \mu}(x - h(c) + n + 2)(x + h(c) + m)$$

$$\bar{P}_{1,\mathbf{q}}(x - \lambda_1) = \bar{c}_{1,\mathbf{q}} \bar{B}_{m,n}(x)$$
(1.6)
$$\times \prod_{c \in \lambda}(x - h(c) + m + 1)(x + h(c) + n) \prod_{c \in \mu}(x - h(c) + n + 1)(x + h(c) + m),$$

where for a cell $c = (i, j)$ of a partition diagram, $h(c)$ is defined to be $i + j$ (in all formulas above empty products are considered equal to 1).

Denote by $\mathbf{l}^{(i)}$ the list obtained from $\mathbf{l}$ by omitting its i-th element, and by $\mathbf{l} - 1$ the list obtained from $\mathbf{l}$ by subtracting 1 from each of its elements (except in case $l_1 = 1$, when $\mathbf{l} - 1$ is defined to consist of $l_2 - 1, \cdots, l_m - 1$).

Our generalization of MacMahon's enumeration of plane partitions contained in a symmetric box is given in Theorem 1.1 below. Its proof is deduced in Section 3 from Proposition 2.1, which interprets the above-defined polynomials $P_{\mathbf{q},1}(x)$ and $\bar{P}_{\mathbf{q},1}(x)$ as tiling enumerators of two families of lattice regions defined in Section 2. Proposition 2.1 and Theorem 1.1 are the main results of this paper.

Let $M(R)$ denote the number of lozenge tilings of the region R (the "M" stands for "matchings," as lozenge tilings can be identified with perfect matchings on the dual graph).

THEOREM 1.1. *The number of lozenge tilings of the above defined regions* $H_1(a, b, k)$, $H_{1,\mathbf{q}}(a, b, k)$, *and* $\bar{H}_{1,\mathbf{q}}(a, b, k)$ *is given by the following formulas:*
(a) For k even, we have

(1.7)
$$2^{-m} M(H_1(a, b, k)) = \begin{cases} P_{\emptyset,1}((a + k - 2)/2)\bar{P}_{1-1,\emptyset}(a/2), & a \text{ even}, l_1 = 1, \\ P_{\emptyset,1}((a + k - 2)/2)P_{1-1,\emptyset}(a/2), & a \text{ even}, l_1 > 1, \\ P_{\emptyset,1^{(m)}}((a + k - 1)/2)\bar{P}_{1,\emptyset}((a - 1)/2), & a \text{ odd}. \end{cases}$$

(b) For k odd, we have

(1.8) $$2^{-m-n} M(H_{1,\mathbf{q}}(a, b, k)) = \begin{cases} \bar{P}_{1,\mathbf{q}}((a + k - 1)/2)P_{\mathbf{q},1^{(m)}}(a/2), & a \text{ even}, \\ \bar{P}_{1,\mathbf{q}^{(n)}}((a + k)/2)P_{\mathbf{q},1}((a - 1)/2), & a \text{ odd}, \end{cases}$$

and

(1.9) $$2^{-m-n} M(\bar{H}_{1,\mathbf{q}}(a, b, k)) = \begin{cases} P_{1,\mathbf{q}}((a + k - 1)/2)\bar{P}_{\mathbf{q},1^{(m)}}(a/2), & a \text{ even}, \\ P_{1,\mathbf{q}^{(n)}}((a + k)/2)\bar{P}_{\mathbf{q},1}((a - 1)/2), & a \text{ odd}. \end{cases}$$

REMARK 1.2. For $k \neq 0$, the bijection between tilings of our regions and stacks of unit cubes breaks down. Indeed, trying to lift such a tiling to three dimensions one is lead to an "impossible" construction, similar to Penrose's impossible triangle (see Figure 1.5).

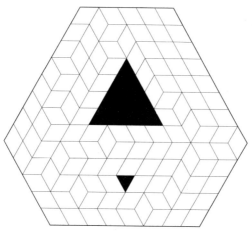

FIGURE 1.5. A tiling of $H_{(1),(1,4,5)}(5,5,3)$.

In Section 3 we show how to reduce the statement of Theorem 1.1 to the problem of finding the tiling generating function of certain simply-connected regions, defined in Section 2. In Sections 4 and 5 we prove, using an inductive argument, that these tiling generating functions are given by the above-defined polynomials P and $\bar{P}$. The idea of the proof is very simple: using combinatorial considerations, we deduce recurrence relations for the tiling generating functions of our regions, and we show that the polynomials P and $\bar{P}$ satisfy the same recurrences. This approach, however, raises the question of how could one guess that the tiling generating functions of our regions are given by the intricate formulas (1.1)–(1.6). We explain in Section 6 how this can be done in a conceptual way, using some computer assistance.

Two families of regions

We now define two families of regions that will turn out to be closely related to the polynomials P and $\bar{P}$ given by (1.5) and (1.6).

Consider the triangular lattice drawn so that one of the directions of the lattice lines is horizontal. From each point of the lattice, there are six possible steps one can take on the lattice. We call the four non-horizontal steps, according to their approximate cardinal direction, northeast, northwest, southwest and southeast.

Let O be a point of the triangular lattice and consider the infinite lattice paths P_u and P_d that start from O and take alternate steps northeast and northwest, respectively southwest and southeast (these paths are shown in dotted lines in Figure 2.1(a)). We say that two consecutive steps of P_u, the first going northeast and the second northwest, form a *bump*; similarly, a step southeast followed by a step southwest on P_d form a bump. Label the bumps of P_u and P_d, starting with the ones closest to O, consecutively by $1, 2, \ldots$. Select arbitrary finite subsets of the bumps of P_d and P_u, and suppose their labels, in increasing order, form (possibly empty) lists $\mathbf{l} = (l_1, \ldots, l_m)$ and $\mathbf{q} = (q_1, \ldots, q_n)$, respectively. Assume that at least one of $\mathbf{l}$ and $\mathbf{q}$ is non-empty.

Starting from these selected bumps, we define the "right boundary" of a region $R_{\mathbf{l},\mathbf{q}}(x)$ as follows (here x is a non-negative integer on which a lower bound will be imposed below). Call the uppermost and lowermost points on a bump the top and bottom of that bump, respectively. For each selected bump on P_u, draw a horizontal ray to the left of its top, and a ray going southwest from its bottom. For each selected bump on P_d, draw a ray going left from its bottom and a ray going northwest from its top.

By following these rays, we join the selected bumps on P_u in a connected piece: follow the horizontal ray from the top of bump q_i to its intersection point with the ray starting from the bottom b of bump q_{i+1}, and then use the latter ray to reach b. The selected bumps on P_d are connected into a single piece in a similar way.

To join the above two pieces together, we draw a horizontal ray left of O. We follow the ray going southwest from the bottom of bump q_1 of P_u until its intersection with the horizontal ray originating at O. Then we follow this latter ray until it hits the ray going northwest from the top a of the l_1-th bump of P_d, which we then track until we reach a. We have now included all selected bumps in a connected path Q joining the top A of the last selected bump on P_u to the bottom B of the last selected bump on P_d (in case $\mathbf{l}$ is empty, B is taken to be the intersection of the ray going southwest from the bottom of the first selected bump on P_u with the horizontal through O; for $\mathbf{q} = \emptyset$, A is chosed analogously; see Figures 2.1(b) and 2.1(c)). It is easy to see that this path has exactly $2l_m - m + n + 1$ steps going southeast (the one exception is the case $\mathbf{l} = \emptyset$, when it has $2l_m - m + n = n$ steps).

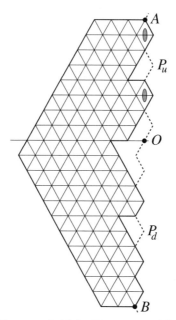

FIGURE 2.1(a). $R_{(2,4,5),(2,4)}(2)$.

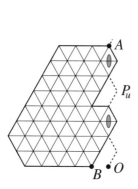

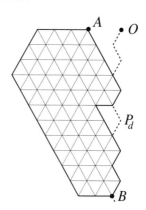

FIGURE 2.1(b). $R_{\emptyset,(2,4)}(4)$.　　　　FIGURE 2.1(c). $R_{(2,4,5),\emptyset}(2)$.

To close the boundary of $R_{\mathbf{l},\mathbf{q}}(x)$, we continue from B by taking x steps left (in case $\mathbf{l}$ is empty, we move left until we are $x+1$ units from O). Then we head northwest for $2l_m - m + n + 1$ steps (one fewer when $\mathbf{l} = \emptyset$), after which we turn and go northeast until we hit the horizontal ray going left from A (see Figure 2.1(a)). Finally, take $x + (l_m - m) - (q_n - n) + 1$ steps east on this ray, until we reach A. It is not hard to see that the length of the southwest portion of this boundary must be chosen as indicated (i.e., to match the number of southeast steps on Q) in order for the enclosed region to admit lozenge tilings (this follows for example by encoding tilings as families of non-intersecting lattice paths, as described at the beginning of Section 4).

In general we will allow the lozenges in a tiling to be weighted. The weight of a tiling is the product of the weights on its lozenges. The sum of weights of all tilings of a region R is denoted $\mathrm{M}(R)$ and is called the tiling generating function of R.

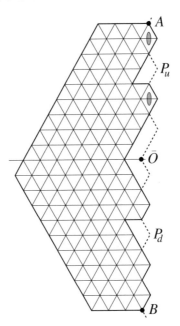

FIGURE 2.2(a). $\bar{R}_{(2,4,5),(2,4)}(3)$.

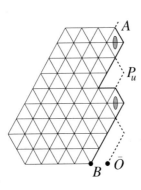

FIGURE 2.2(b). $\bar{R}_{\emptyset,(2,4)}(5)$.

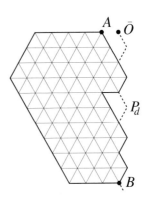

FIGURE 2.2(c). $\bar{R}_{(2,4,5),\emptyset}(3)$.

We define $R_{\mathbf{l},\mathbf{q}}(x)$ to be the region enclosed by the above-defined boundary, requiring in addition that the tile positions fitting in the selected bumps on P_u are weighted by $1/2$ (i.e., if one of these positions is occupied by a lozenge in some tiling, that lozenge gets weight $1/2$ — this will be indicated in our figures by a shaded oval placed on such tile positions; all other tiles get weight 1). Since the top and bottom sides have non-negative length, x must satisfy $x \geq \max\{0, q_n - l_m - n + m - 1\}$ (and $x \geq \max\{-1, q_n - n - 1\} = q_n - n - 1$, for $\mathbf{l} = \emptyset$). Define $R_{\emptyset,\emptyset}(x) = \emptyset$ for all x.

Our second family of regions, denoted $\bar{R}_{\mathbf{l},\mathbf{q}}(x)$, is defined almost identically (see Figures 2.2(a)–(c)). The only difference is that, in the construction of the path Q that connects the selected bumps, instead of the horizontal ray starting from O, we use the one starting from the lattice point $\bar{O}$, one step southwest of

O (and, in case $\mathbf{l}$ is empty, when defining the bottom side of $\bar{R}_{\emptyset,\mathbf{q}}(x)$, we move left until we are x units away from $\bar{O}$). As a consequence, this path has now one fewer southeast steps than before, and hence we define the southwest boundary of $\bar{R}_{\mathbf{l},\mathbf{q}}(x)$ to have length $2l_m - m + n$. Also, the top side of the enclosed region has now length $x + (l_m - m) - (q_n - n)$. The parameter x satisfies therefore $x \geq \max\{0, q_n - l_m - n + m\}$ (this time the condition on x remains the same for $\mathbf{l} = \emptyset$). Again, we define $\bar{R}_{\emptyset,\emptyset}(x) = \emptyset$ for all x.

The close connection between the above-defined regions and the polynomials P and $\bar{P}$ given by (1.5) and (1.6) is expressed by the following result.

PROPOSITION 2.1. *For all lists $\mathbf{l}$ and $\mathbf{q}$ of strictly increasing positive integers we have*

(2.1) $$\mathrm{M}(R_{\mathbf{l},\mathbf{q}}(x)) = P_{\mathbf{l},\mathbf{q}}(x)$$
(2.2) $$\mathrm{M}(\bar{R}_{\mathbf{l},\mathbf{q}}(x)) = \bar{P}_{\mathbf{l},\mathbf{q}}(x),$$

for all x for which the regions on the left hand side of these equalities are defined.

In the next section we show how the above Proposition implies Theorem 1.1. The proof of Proposition 2.1 is presented in Sections 4 and 5.

Reduction to simply-connected regions

One useful way to approach certain tiling enumeration problems (lozenge tilings in particular) is to identify them with families of non-intersecting lattice paths, and then use the Gessel-Viennot theorem (see e.g. [**12**, Theorem 1.2] or [**5**]) to express the number we seek as a determinant. While this could be carried out directly for the regions on the left hand sides of (1.7)–(1.9) that we are concerned with, it turns out that it will be more effective to reduce first our problem to the case of certain simply connected subregions, for the number of tilings of which we can easily obtain recurrences. This reduction can be achieved using the Factorization Theorem for perfect matchings presented in [**1**, Theorem 1.2].

Let R be a region of the triangular lattice, and suppose R has a vertical symmetry axis ℓ (an example is shown in Figure 3.1). Let P_l and P_r be the two zig-zag lattice paths taking alternate steps northeast and northwest and touching ℓ from left and right, respectively. Group the unit triangles of R crossed by ℓ in sequences of contiguous triangles. These are separated by sequences of triangles of the lattice not contained in R; we call these sequences gaps.

We cut R into two subregions R^+ and R^- as follows. Consider the paths P_l and P_r. Start at top by following P_r to the first gap. After each gap, keep following the path along which we reached it or switch to the other path, according as the gap contains an even or an odd number of unit triangles. Continue this until we reach the bottom of R. Finally, assign weight $1/2$ to all tile positions in R that have two sides contained in our cutting path. Define R^+ and R^- to be the regions obtained to the left and right of our cutting path, respectively, and having the above-mentioned lozenge positions weighted by $1/2$.

Since we are interested in counting lozenge tilings, we may assume that R has an even number of unit triangles crossed by ℓ (indeed, if this number was odd, the symmetry of R would imply that R contains an odd number of unit triangles, so it would have no lozenge tilings); define the *width* of R, denoted $\mathrm{w}(R)$, to be half this number.

Then the Factorization Theorem for perfect matchings of [**1**, Theorem 1.2] implies

$$(3.1) \qquad \mathrm{M}(R) = 2^{\mathrm{w}(R)} \, \mathrm{M}(R^+) \, \mathrm{M}(R^-).$$

Indeed, lozenge tilings of R can be identified with perfect matchings of the dual graph G of R, i.e., the graph whose vertices are the unit triangles of R and whose edges connect precisely those pairs of triangles that share an edge. Since R is symmetric with respect to ℓ, so is G, and all conditions in the hypothesis of [**1**, Theorem 1.2] are easily seen to be met. We obtain that $\mathrm{M}(G) = 2^k \, \mathrm{M}(G^+) \, \mathrm{M}(G^-)$, where G^+ and G^- are certain precisely defined subgraphs, and $\mathrm{M}(G)$ denotes the matching

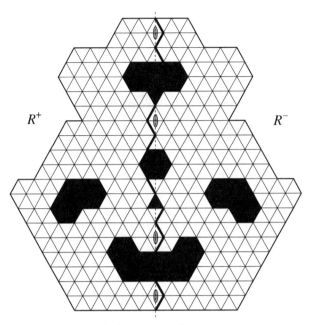

FIGURE 3.1.

generating function of G. However, it is easily checked that G^+ and G^- are exactly the dual graphs of the regions R^+ and R^-, respectively, thus proving (2.1).

PROOF OF THEOREM 1.1. All equalities in the statement of Theorem 1.1 follow directly by applying (2.1) to the regions on the left hand side of (1.7)–(1.9) and using Proposition 2.1. More precisely, in all cases, the absolute value of the exponent in the power of 2 in (1.7)–(1.9) turns out to be the width of the corresponding region, and, by Proposition 2.1, the two factors in the product on the right hand side turn out to be equal to the tiling generating functions of the two regions that arise by applying the Factorization Theorem to our regions.

(a) Take R to be the region $H_1(a, b, k)$, and apply (2.1). Since all the gaps are even, the cutting path determining R^+ and R^- lies fully on the right side of ℓ (this is illustrated in Figures 3.2(a)–(c)). Suppose a is even. Since $a + k$ is also even, this cutting path starts at the midpoint of the top side of R and ends at the midpoint of its base (see Figures 3.2(a) and (b)). Consequently, all unit triangles of R crossed by ℓ can be paired to form rhombic vertebrae. By our definition, these are precisely the l_1-th, ..., l_m-th vertebrae of the hexagon H enclosed by the outer boundary of R. Therefore, the region R^+ to the left of the cutting path is precisely $R_{\emptyset,1}((a + k - 2)/2)$.

As for the region R^- to the right of the cutting path, suppose first that $l_1 > 1$ (see Figure 3.2(a)). Then it is easy to check that the region obtained from R^- after removing the forced lozenges is precisely $\bar{R}_{(l_1-1,...,l_m-1)}(a/2)$, rotated by 180 degrees. On the other hand, if $l_1 = 1$, because of the different pattern of forced lozenges in R^- (extra forced lozenges on the bottom, as shown in Figure 3.2(b)), the subregion left after removing the forced lozenges is isomorphic to $R_{(l_2-1,...,l_m-1)}(a/2)$. Since the width of R is m, this proves, by Proposition 2.1, the first two equalities in (1.7).

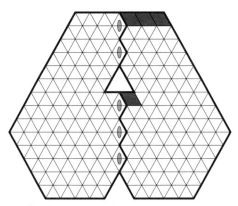

FIGURE 3.2(a).
$H_{(2,3,4,6,7)}(6,5,4)$ reduces to
$R_{\emptyset,(2,3,4,6,7)}(4)$ and $R_{(1,2,3,5,6),\emptyset}(3)$.

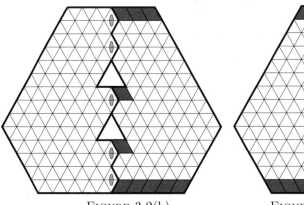

FIGURE 3.2(b).   FIGURE 3.2(c).
$H_{(1,2,4,6,7)}(6,5,4)$ reduces to $H_{(1,3,4,6,7)}(5,5,4)$ reduces to
$R_{\emptyset,(1,2,4,6,7)}(4)$ and $\bar{R}_{(1,3,5,6),\emptyset}(3)$. $R_{\emptyset,(1,3,4,6)}(4)$ and $\bar{R}_{(1,3,4,6,7),\emptyset}(2)$.

For a odd we proceed similarly. Now the last labeled vertebra of H is a triangular vertebra, and it remains present for all choices of removing even Δ-windows from H (see Figure 3.2(c)). By definition, its label is l_m. This fact and an examination of the pattern of forced lozenges in R^+ and R^- explains why, at the indices of the two factors on the right hand side of the third equality in (1.7), the list $\mathbf{l}$ appears once with its last entry omitted and once in full. Indeed, because of forced lozenges at the top of R^+, the triangular vertebra labeled l_m is not relevant for the region obtained from R^+ by removing the forced lozenges. This leftover region is easily seen to be exactly $\bar{R}_{\emptyset,\mathbf{l}^{(m)}}((a+k-1)/2)$. Furthermore, the region obtained from R^- by removing the forced lozenges is isomorphic to $\bar{R}_{\mathbf{l},\emptyset}((a-1)/2)$, irrespective of the value of l_1 (the difference in this regard from the case a even is explained by the fact that the pattern of forced lozenges in R^- depends on l_1 for a even, but does not for a odd). Since the width of R is m, by Proposition 2.1 this completes the proof of (1.7).

(b) To prove (1.8), take R to be the region $H_{\mathbf{l},\mathbf{q}}(a,b,k)$ and apply (2.1). The resulting cutting path stays on the right of ℓ until it arrives to the odd Δ-window,

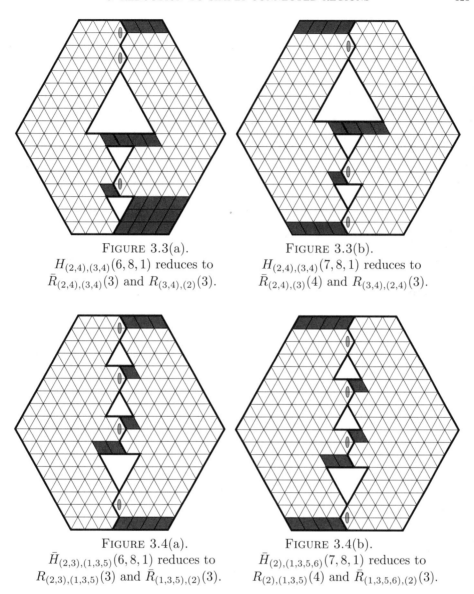

FIGURE 3.3(a).
$H_{(2,4),(3,4)}(6,8,1)$ reduces to
$\bar{R}_{(2,4),(3,4)}(3)$ and $R_{(3,4),(2)}(3)$.

FIGURE 3.3(b).
$H_{(2,4),(3,4)}(7,8,1)$ reduces to
$\bar{R}_{(2,4),(3)}(4)$ and $R_{(3,4),(2,4)}(3)$.

FIGURE 3.4(a).
$\bar{H}_{(2,3),(1,3,5)}(6,8,1)$ reduces to
$R_{(2,3),(1,3,5)}(3)$ and $\bar{R}_{(1,3,5),(2)}(3)$.

FIGURE 3.4(b).
$\bar{H}_{(2),(1,3,5,6)}(7,8,1)$ reduces to
$R_{(2),(1,3,5)}(4)$ and $\bar{R}_{(1,3,5,6),(2)}(3)$.

then it switches to the left of ℓ and stays there until it reaches the bottom of R (this is illustrated in Figures 3.3(a) and (b)).

Recall that our reference line for labeling the vertebrae of H is now the base of the odd removed window; denote it by $\mathcal{R}$. Suppose a is even. Then the last vertebra below $\mathcal{R}$ is in this case a triangular vertebra, which remains present for all choices of removed windows (see Figure 3.3(a)). The rest of the unit triangles of R crossed by ℓ can be paired to form rhombic vertebrae. By definition, the region R contains precisely the vertebrae labeled $l_1, \ldots, l_m$ below $\mathcal{R}$, and those labeled $q_1, \ldots, q_n$ above $\mathcal{R}$. It is easy to check that the region obtained from R^+ after removing the forced lozenges is precisely $\bar{R}_{\mathbf{l},\mathbf{q}}((a+k-1)/2)$, while the one obtained from R^- after the same procedure is congruent to $R_{\mathbf{q},\mathbf{l}^{(m)}}(a/2)$. Since the width of R is $m+n$, we obtain by Proposition 2.1 the first equality of (1.8).

If a is odd, a similar analysis shows that R^+ is congruent to $\bar{R}_{\mathbf{1},\mathbf{q}^{(n)}}((a+k)/2)$, and R^- to $R_{\mathbf{q},\mathbf{1}}((a-1)/2)$ (see Figure 3.3(b) for an illustration). Since the width of R is again $m+n$, Proposition 2.1 implies the second equality of (1.8).

The regions $\bar{H}_{\mathbf{1},\mathbf{q}}(a,b,k)$ are treated in a perfectly similar way (this is illustrated in Figures 3.4(a) and (b)). Again, the regions R^+ and R^- obtained by applying (2.1) turn out to be members of the families of regions defined in Section 2, and applying Proposition 2.1 one arrives at the formulas (1.9). $\square$

REMARK 3.1. Part (a) of Theorem 1.1 (i.e., the case k even) also follows from Lemma 2.2 of [6]. Indeed, as shown in the next section, it is easy to express the tiling generating functions of the regions R^+ and R^- arising by applying the Factorization Theorem to $H_{\mathbf{1}}(a,b,k)$ as determinants of Gessel-Viennot matrices (see (4.1)). It is also not hard to check that the entries of these matrices are of the form required by [6, Lemma 2.2], which therefore provides explicit product formulas for the value of these determinants.

For k odd on the other hand, the corresponding Gessel-Viennot matrices have a more complicated structure and the evaluation of their determinants does not follow from Lemma 2.2 of [6].

Recurrences for $\mathrm{M}(R_{\mathbf{l},\mathbf{q}}(x))$ and $\mathrm{M}(\bar{R}_{\mathbf{l},\mathbf{q}}(x))$

Let $\mathbf{l} = (l_1, \ldots, l_m)$ and $\mathbf{q} = (q_1, \ldots, q_n)$ be two (possibly empty) lists of strictly increasing positive integers, and let $\mathcal{T}$ be a tiling of $R_{\mathbf{l},\mathbf{q}}(x)$ (Figure 4.1 shows an example for $\mathbf{l} = (2, 3)$, $\mathbf{q} = (2, 4)$ and $x = 3$; a white center on a tile indicates that it is weighted by $1/2$). This tiling divides the southwestern boundary of $R_{\mathbf{l},\mathbf{q}}(x)$ into $2l_m - m + n + 1$ unit segments. Starting from these segments, by following the tiles of $\mathcal{T}$, one obtains paths of rhombi that end on the southwest-facing unit segments on the right boundary of $R_{\mathbf{l},\mathbf{q}}(x)$ (these paths are shaded dark in Figure 4.1). It is clear that these paths are non-intersecting, and it is not hard to see that they completely determine $\mathcal{T}$. Since along the paths of rhombi the only possible steps are northeast and east, it is clear that they can be identified with lattice paths on the square lattice $\mathbb{Z}^2$ taking steps north and east.

We will need the following special case of the Gessel-Viennot theorem for non-intersecting lattice paths (see [5]). All our lattice paths will be on the grid graph $\mathbb{Z}^2$, directed so that its edges point east and north. We allow the edges of $\mathbb{Z}^2$ to be weighted, and define the weight of a lattice path to be the product of the weights on its steps. The weight of an N-tuple of lattice paths is the product of the individual weights of its members. The generating function of a set of N-tuples of lattice paths is the sum of the weights of its elements.

An N-tuple of starting points $\mathbf{u} = (u_1, \ldots, u_N)$ is said to be *compatible* with an N-tuple of ending points $\mathbf{v} = (v_1, \ldots, v_N)$ if for all $i < j$ and $k < l$, every lattice path from u_i to v_l intersects each lattice path from u_j to v_k.

THEOREM 4.1 (GESSEL-VIENNOT). *Let $\mathbf{u}$ and $\mathbf{v}$ be compatible N-tuples of starting and ending points on the graph $\mathbb{Z}^2$ oriented as above. Then the generating function for N-tuples of non-intersecting lattice paths joining the starting points to the ending points is*

$$\det\left((a_{ij})_{1 \le i,j \le N}\right),$$

where a_{ij} is the generating function for lattice paths from u_i to v_j.

We will find it convenient to view the paths of rhombi directly as lattice paths on $\mathbb{Z}^2$, thus bypassing the "extra steps" of bijecting them with lattice paths on a lattice of rhombi with angles of 60 and 120 degrees, and then deforming this to the square lattice. In this context, the "points" of our lattice $\mathcal{L}$ are the southwest-facing edges of the triangular lattice — we call them *segments* —, and the "lines" of $\mathcal{L}$ are sequences of adjacent rhombi extending horizontally or in the southwest-northeast direction — we call these *rows* and *columns*, respectively. The rhombi that make up these rows and columns are called *edges* of $\mathcal{L}$.

To coordinatize our lattice, we choose the x-axis to be the bottommost row, and the y-axis to be the leftmost column, intersecting $R_{\mathbf{l},\mathbf{q}}(x)$. Let $N = 2l_m - m + n + 1$,

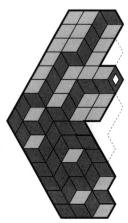

FIGURE 4.1. Encoding a tiling as paths of rhombi.

and label the starting and ending segments of the lattice paths encoding our tiling $\mathcal{T}$, from bottom to top, by $u_1, \ldots, u_N$ and $v_1, \ldots, v_N$, respectively.

Weight by $1/2$ the edges of $\mathcal{L}$ corresponding to tile positions weighted by $1/2$ in $R_{\mathbf{l},\mathbf{q}}(x)$. Weight all other edges of $\mathcal{L}$ inside $R_{\mathbf{l},\mathbf{q}}(x)$ by 1, and the edges outside $R_{\mathbf{l},\mathbf{q}}(x)$ by 0. By Theorem 4.1 and the above-mentioned bijection between tilings and lattice paths, we obtain that

$$(4.1) \qquad \mathrm{M}(R_{\mathbf{l},\mathbf{q}}(x)) = \det\left((a_{ij})_{1 \le i,j \le N}\right),$$

where a_{ij} is the generating function for lattice paths from u_i to v_j.

Suppose $m \le n$. Then expanding this determinant along the last row turns out to give us a recurrence relation for $\mathrm{M}(R_{\mathbf{l},\mathbf{q}}(x))$.

LEMMA 4.2. *For $m \le n$, we have*

$$(4.2) \qquad \mathrm{M}(R_{\mathbf{l},\mathbf{q}}(x)) = \sum_{k=1}^{n}(-1)^{n-k}C_k\,\mathrm{M}(R_{\mathbf{l},\mathbf{q}^{(k)}}(x)) + (-1)^n \delta_{mn}\,\mathrm{M}(\bar{R}_{\mathbf{l},\mathbf{q}}(x)),$$

where δ_{mn} is the Kronecker symbol and

$$(4.3) \qquad C_k = \binom{x + l_m + q_k}{2q_k + m - n} + \frac{1}{2}\binom{x + l_m + q_k}{2q_k + m - n - 1}.$$

PROOF. Let A be the matrix in (4.1). Expanding the determinant in (4.1) along the last row we obtain

$$(4.4) \qquad \mathrm{M}(R_{\mathbf{l},\mathbf{q}}(x)) = \sum_{k=0}^{N-1}(-1)^k a_{N,N-k}\det(A^{N,N-k}),$$

where A^{ij} is the matrix obtained from A by deleting row i and column j.

We claim that at most the first $n + 1$ terms in the sum (4.4) are nonzero. Indeed, recall that $a_{N,N-k}$ is the generating function for lattice paths on $\mathcal{L}$ going from the segment u_N to the segment v_{N-k}. It follows from our construction of

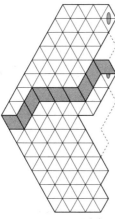
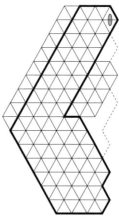

FIGURE 4.2(a).   FIGURE 4.2(b).

$R_{\mathbf{l},\mathbf{q}}(x)$ that there is a unique segment s on its right boundary contained in the row immediately below O, and s is facing southeast. The y-coordinates of the segments u_N and s work out to be, in our coordinatization, $2l_m - m + n$ and $2l_m$, respectively. Since our paths on $\mathcal{L}$ take steps northeast and east, and $m \leq n$, we see that a path starting at u_N will end, on the right boundary of $R_{\mathbf{l},\mathbf{q}}(x)$, at some segment above s, or, possibly, in case $m = n$, at s itself (see Figures 4.2(a), 4.3(a) and 4.4(a)). Since there are precisely n segments above s on the right boundary of $R_{\mathbf{l},\mathbf{q}}(x)$ (one corresponding to each selected bump on P_u), this proves our claim (and also implies $s = v_{N-n}$).

Furthermore, it is easy to determine the exact value of $a_{N,N-k}$ for $k \leq n$. It follows from the previous paragraph that $a_{N,N-n} = 0$ unless $m = n$, in which case $a_{N,N-n} = 1$ (this accounts for the Kronecker symbol in (4.2)). For $1 \leq k \leq n$, $a_{N,N-n+k}$ is the generating function of lattice paths going from u_N to v_{N-n+k}. In our system of coordinates it turns out that we have

$$u_N = (0, 2l_m - m + n)$$
$$v_{N-n+k} = (x + l_m - q_k - m + n + 1, 2q_k + 2l_m), \quad \text{for } 1 \leq k \leq n.$$

Partitioning the lattice paths from u_N to v_{N-n+k} in two classes according to the direction of their last step we obtain that $a_{N,N-n+k} = C_k$, with C_k given by (4.3). Replacing the upper summation limit in (4.4) by n and then replacing the summation index k by $n - k$ we obtain from (4.4) that

$$(4.5) \quad \mathrm{M}(R_{\mathbf{l},\mathbf{q}}(x)) = \sum_{k=1}^{n} (-1)^{n-k} C_k \det(A^{N,N-n+k}) + (-1)^n \delta_{mn} \det(A^{N,N-n}).$$

To complete the proof of the Lemma, we show that the determinants on the right hand side of (4.5) are equal to the corresponding tiling generating functions on the right hand side of (4.2).

More precisely, we claim that, for $1 \leq k \leq n$, we have $\det(A^{N,N-n+k}) = \mathrm{M}(R_{\mathbf{l},\mathbf{q}^{(k)}}(x))$. Indeed, by Theorem 4.1 it follows that $\det(A^{N,N-n+k})$ is equal to the generating function of $(N-1)$-tuples of non-intersecting lattice paths starting at $u_1, \ldots, u_{N-1}$ and ending at $v_1, \ldots, v_{N-n+k-1}, v_{N-n+k+1}, \ldots, v_N$. However, by the

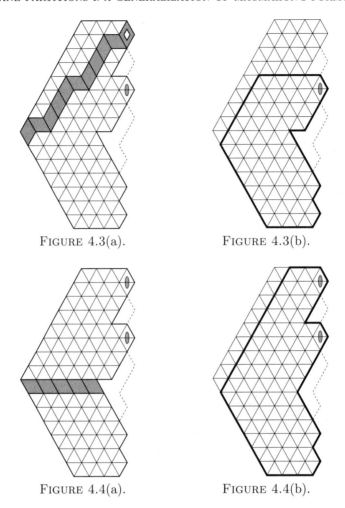

FIGURE 4.3(a).   FIGURE 4.3(b).

FIGURE 4.4(a).   FIGURE 4.4(b).

bijection between tilings and lattice paths, this is precisely the generating function for tilings of $R_{\mathbf{l},\mathbf{q}^{(k)}}(x)$ (for $\mathbf{l} = (2,3)$, $\mathbf{q} = (2,4)$, this is illustrated in Figures 4.2 and 4.3; the starting and ending segments of the paths on the left are omitted, leading to the regions on the right).

Similarly, one sees that $\det(A^{N,N-n}) = \mathrm{M}(\bar{R}_{\mathbf{l},\mathbf{q}}(x))$. Indeed, $\det(A^{N,N-n})$ is the generating function for $(N-1)$-tuples of non-intersecting lattice paths going from $u_1, \ldots, u_{N-1}$ to $v_1, \ldots, v_{N-n-1}, v_{N-n+1}, \ldots, v_N$, respectively. In this case the corresponding region has the same sets of selected bumps as $R_{\mathbf{l},\mathbf{q}}(x)$, but its boundary around O is changed so that it becomes precisely $\bar{R}_{\mathbf{l},\mathbf{q}}(x))$ (see Figures 4.4(a) and (b)). This completes the proof of the Lemma. $\square$

The above result gives a recurrence relation for $\mathrm{M}(R_{\mathbf{l},\mathbf{q}}(x))$ when $m \leq n$. Note that this assumption was indeed necessary: for $m > n$ there are additional non-zero terms in the sum (4.4), and the regions they can be associated to are *not* part of our families R and $\bar{R}$. To deal with the case $m > n$ we use the "symmetry" of $R_{\mathbf{l},\mathbf{q}}(x)$: we encode its tilings by non-intersecting lattice paths starting on its northwestern side and taking steps southeast and east. This leads us to the following result.

LEMMA 4.3. *For $m > n$, we have*

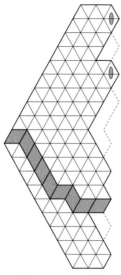 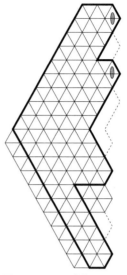

FIGURE 4.5(a).   FIGURE 4.5(b).

(4.6)

$$\mathrm{M}(R_{\mathbf{l},\mathbf{q}}(x)) = \sum_{k=1}^{m-1} (-1)^{m-k} D_k \,\mathrm{M}(R_{\mathbf{l}^{(k)},\mathbf{q}}(x-1)) + D_m \,\mathrm{M}(R_{\mathbf{l}^{(m)},\mathbf{q}}(x+l_m-l_{m-1}-1)),$$

where

(4.7)
$$D_k = \binom{x+l_m+l_k-m+n+1}{2l_k-m+n+1}.$$

PROOF. We proceed in the same fashion as in the proof of Lemma 4.2. First, we change the definition of the lattice $\mathcal{L}$ so that its points (called again segments) are the unit edges of the triangular lattice facing northwest, and the lines (still called rows and columns) are sequences of adjacent unit rhombi extending horizontally and in the northwest-southeast direction. We choose the x- and y-axis to be the topmost row and the leftmost column intersecting $R_{\mathbf{l},\mathbf{q}}(x)$, respectively. Since the northwestern side of $R_{\mathbf{l},\mathbf{q}}(x)$ has length $2q_n+m-n$, a tiling of $R_{\mathbf{l},\mathbf{q}}(x)$ is encoded by this many non-intersecting lattice paths in $\mathcal{L}$. Let $N = 2q_n+m-n$ and label the starting and ending segments of these lattice paths, from top to bottom, by $u_1,\ldots,u_N$ and $v_1,\ldots,v_N$, respectively.

Formula (4.1) holds also with this new definitions of $\mathcal{L}$, u_i and v_i. Expanding the determinant on its right hand side along the last row we obtain

(4.8)
$$\mathrm{M}(R_{\mathbf{l},\mathbf{q}}(x)) = \sum_{k=0}^{N-1} (-1)^k a_{N,N-k} \det(A^{N,N-k}),$$

where the entries of A are now defined in terms of the new u_i's and v_i's.

We claim that all terms in this sum except for the first m are 0. Indeed, denote by s the unique segment in the right boundary of $R_{\mathbf{l},\mathbf{q}}(x)$ contained in the row

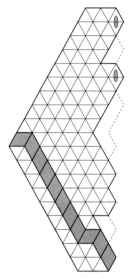
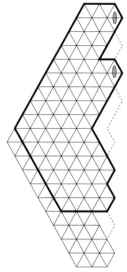

FIGURE 4.6(a).   FIGURE 4.6(b).

immediately above O. The y-coordinates of u_N and s in $\mathcal{L}$ are $2q_n + m - n - 1$ and $2q_n - 1$, respectively. Since $m > n$, and our paths take steps southeast and east, it follows that the path starting at u_N ends at a segment below s on the right boundary of $R_{\mathbf{l},\mathbf{q}}(x)$ (this is illustrated in Figures 4.5(a) and 4.6(a) for $\mathbf{l} = (2,3,5)$, $\mathbf{q} = (2,4)$ and $x = 2$). There are exactly m such segments, one for each selected bump on P_d. So $a_{Nk} = 0$ unless $k \in \{N - m + 1, \ldots, N\}$, which proves our claim.

For $1 \le k \le m$, the exact value of $a_{N,N-m+k}$ follows from the coordinates of u_N and v_{N-m+k}. Since we have

$$u_N = (0, 2q_n + m - n - 1)$$
$$v_{N-m+k} = (x + l_m - l_k, 2l_k + 2q_n), \quad \text{for } 1 \le k \le m,$$

we obtain that $a_{N,N-m+k} = D_k$, where D_k is given by (4.7). Therefore, replacing the upper summation limit in (4.8) by $m - 1$ and then replacing the summation index k by $m - k$, (4.8) becomes

$$(4.9) \qquad \mathrm{M}(R_{\mathbf{l},\mathbf{q}}(x)) = \sum_{k=1}^{m-1} (-1)^{m-k} D_k \det(A^{N,N-m+k}) + D_m \det(A^{NN}).$$

To complete the proof of the Lemma, we show that the determinants on the right hand side of (4.9) are equal to the corresponding tiling generating functions on the right hand side of (4.6).

We claim that, for $1 \le k \le m-1$, we have $\det(A^{N,N-m+k}) = \mathrm{M}(R_{\mathbf{l}^{(k)},\mathbf{q}}(x-1))$. Indeed, by Theorem 4.1, $\det(A^{N,N-m+k})$ is equal to the tiling generating function for $(N-1)$-tuples of non-intersecting lattice paths in $\mathcal{L}$ starting at $u_1, \ldots, u_{N-1}$ and ending at $v_1, \ldots, v_{N-m+k-1}, v_{N-m+k+1}, \ldots, v_N$. By the bijection between tilings and lattice paths, this is precisely the generating function for tilings of $R_{\mathbf{l}^{(k)},\mathbf{q}}(x-1)$. (The change in the parameter x is a new ingredient that was not present in the case $m \le n$; this is due to the change of our coordinate system in the present case. Figure 4.5 illustrates this for $k = 2$.)

Similarly, one sees that $\det(A^{N,N}) = \mathrm{M}(R_{l^{(m)},\mathbf{q}}(x + l_m - l_{m-1} - 1))$. Indeed, $\det(A^{N,N})$ is the generating function for $(N-1)$-tuples of non-intersecting lattice paths going from $u_1, \ldots, u_{N-1}$ to $v_1, \ldots, v_{N-1}$, respectively. The corresponding region is the portion of $R_{l,\mathbf{q}}(x)$ consisting of the unit triangles in and above the row containing v_{N-1}, and to the right of the column containing u_N (see Figure 4.6). This is readily seen to be exactly $R_{l^{(m)},\mathbf{q}}(x + l_m - l_{m-1} - 1))$, thus completing the proof.   □

We will also need the analogs of the above two Lemmas for the regions $\bar{R}_{l,\mathbf{q}}(x)$.

LEMMA 4.4. *For $m < n$, we have*

$$(4.10) \qquad \mathrm{M}(\bar{R}_{l,\mathbf{q}}(x)) = \sum_{k=1}^{n} (-1)^{n-k} \bar{C}_k \, \mathrm{M}(\bar{R}_{l,\mathbf{q}^{(k)}}(x)),$$

where

$$(4.11) \qquad \bar{C}_k = \binom{x + l_m + q_k}{2q_k + m - n + 1} + \frac{1}{2}\binom{x + l_m + q_k}{2q_k + m - n}.$$

PROOF. Let $\mathcal{L}$ be as in the proof of Lemma 4.2. The southwestern side of $\bar{R}_{l,\mathbf{q}}(x)$ has length $N = 2l_m - m + n$. Encode the tilings of $\bar{R}_{l,\mathbf{q}}(x)$ as N-tuples of non-intersecting lattice paths on $\mathcal{L}$. Order their starting and ending points from bottom to top and express their number as a determinant by Theorem 4.1. Using the same reasoning as in the proof of Lemma 4.2, it follows that only the last n entries in the last row of this determinant are nonzero. These are easily seen to be equal to the numbers $\bar{C}_k$ given by (4.11). Expand this determinant along the last row. Using arguments similar to those in the proof of Lemma 4.2, the order $N - 1$ minors in the determinant expansion can be interpreted as tiling generating functions of regions from the $\bar{R}$-family, and one obtains (4.10).   □

LEMMA 4.5. *For $m \ge n$, we have*

$$\mathrm{M}(\bar{R}_{l,\mathbf{q}}(x)) = \sum_{k=1}^{m-1} (-1)^{m-k} \bar{D}_k \, \mathrm{M}(\bar{R}_{l^{(k)},\mathbf{q}}(x-1))$$

$$(4.12)$$

$$+ \bar{D}_m \, \mathrm{M}(\bar{R}_{l^{(m)},\mathbf{q}}(x + l_m - l_{m-1} - 1)) + (-1)^m \delta_{mn} \, \mathrm{M}(R_{l,\mathbf{q}}(x-1)),$$

where δ_{mn} is the Kronecker symbol and

$$(4.13) \qquad \bar{D}_k = \binom{x + l_m + l_k - m + n}{2l_k - m + n}.$$

PROOF. Choose $\mathcal{L}$ to be as in the proof of Lemma 4.3. The northwestern side of $\bar{R}_{l,\mathbf{q}}(x)$ has length $N = 2q_n + m - n + 1$. Encode the tilings of $\bar{R}_{l,\mathbf{q}}(x)$ as N-tuples of non-intersecting lattice paths on $\mathcal{L}$. Order their starting and ending points from bottom to top and express their number as a determinant by Theorem 4.1. By arguments similar to those in the proof of Lemma 4.3, it follows that only the last

$m + \delta_{mn}$ terms in the last row of this determinant are nonzero. More precisely, it turns out that the $(m+1)$-st to last entry in the last row is δ_{mn}, and the subsequent entries are given by (4.13). Expand the determinant along the last row. In analogy to the situation in the proof of Lemma 4.3, it turns out that the minors of order $N - 1$ in this expansion are equal to tiling generating functions of regions belonging to the $\bar{R}$- and R-families. Precise identification of these regions leads to (4.12). (Note that, except for the last term on the right hand side of (4.12), the shift in the parameter x is precisely the same as in the case of Lemma 4.3. The last term does not have a correspondent in (4.6) because there we didn't allow $m = n$.) □

Finally, to successfully carry out the inductive evaluation of our tiling generating functions, we need to pay attention to their behaviour when $\mathbf{l}$, $\mathbf{q}$ and x are on the "boundary" (i.e., $\mathbf{l}$ or $\mathbf{q}$ is empty, or x takes on the minimum value for which our regions are defined). By definition, we take $M(R_{\emptyset,\emptyset}(x)) = 1$ and $M(\bar{R}_{\emptyset,\emptyset}(x)) = 1$, for all x (empty regions have precisely one tiling).

In the recurrences (4.2), (4.6), (4.10) and (4.12), the value of the argument in the terms on the right hand side is at least as large as one less the value of the argument on the left hand side. Therefore, if at least one of $\mathbf{l}$ and $\mathbf{q}$ is nonempty, these recurrences can be applied as long as the argument x on their left hand side is not minimum possible — i.e., for (4.2) and (4.6), if x is strictly larger than $\max\{0, q_n - l_m - n + m - 1\}$ ($\max\{-1, q_n - n - 1\} = q_n - n - 1$, in case $\mathbf{l} = \emptyset$), and, for (4.10) and (4.12), if x is strictly larger than $\max\{0, q_n - l_m - n + m\}$.

When x does take on its minimum possible value, we can express the tiling generating functions of our regions as shown in the following two Lemmas.

LEMMA 4.6. (a) Suppose $l_m - m + 1 \geq q_n - n$. Then for $\mathbf{l} \neq \emptyset$ we have

$$(4.14) \qquad M(R_{\mathbf{l},\mathbf{q}}(0)) = M(R_{\mathbf{l}^{(m)},\mathbf{q}}(l_m - l_{m-1} - 1)).$$

(b) Let $l_m - m + 1 \leq q_n - n$. Then we have

$$(4.15) \qquad M(R_{\mathbf{l},\mathbf{q}}(q_n - l_m - n + m - 1)) = \frac{1}{2} M(R_{\mathbf{l},\mathbf{q}^{(n)}}(q_n - l_m - n + m - 1)).$$

PROOF. Recall that, for $\mathbf{l} \neq \emptyset$, the lengths of the base and top side of $R_{\mathbf{l},\mathbf{q}}(x)$ are x and $x + (l_m - m) - (q_n - n) + 1$, respectively. The assumption in (a) implies therefore that the base is at most as long as the top side. Thus, by setting $x = 0$ we obtain a valid region, which has all the lozenges along its southwestern side forced (see Figure 4.7(a) for an illustration). The leftover region is readily identified to be $R_{\mathbf{l}^{(m)},\mathbf{q}}(l_m - l_{m-1} - 1)$.

Consider now part (b). Our assumption implies that the top side is shorter than or equal to the base. Indeed, for $\mathbf{l} \neq \emptyset$ this follows from the above paragraph. For $\mathbf{l} = \emptyset$, the top side has length $x - (q_n - n) + 1$, and the base has length $x - q_1 + 2$. Therefore, it follows that the top side is at most as long as the base also in this case.

By setting $x = q_n - l_m - n + m - 1$ the top side shrinks to a point and we obtain a valid region whose lozenges along the northwestern side are forced (see Figure 4.7(b)). The leftover region is easily seen to be $R_{\mathbf{l},\mathbf{q}^{(n)}}(q_n - l_m - n + m - 1)$. Since the top forced tile has weight $1/2$ and the others have weight 1, we obtain (4.15). □

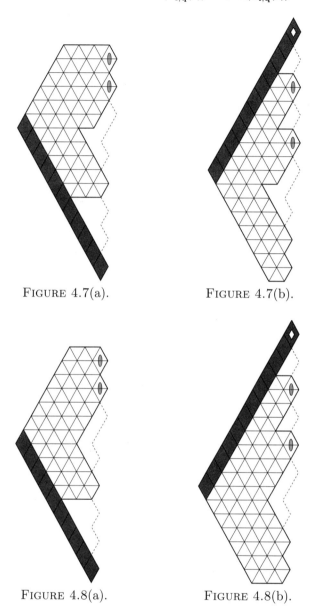

FIGURE 4.7(a).   FIGURE 4.7(b).

FIGURE 4.8(a).   FIGURE 4.8(b).

LEMMA 4.7. (a) *Suppose* $l_m - m \geq q_n - n$. *Then for* $\mathbf{l} \neq \emptyset$ *we have*

$$(4.16) \qquad \mathrm{M}(\bar{R}_{\mathbf{l},\mathbf{q}}(0)) = \mathrm{M}(\bar{R}_{\mathbf{l}^{(m)},\mathbf{q}}(l_m - l_{m-1} - 1)).$$

(b) *Suppose* $l_m - m \leq q_n - n$. *Then for* $\mathbf{q} \neq \emptyset$ *we have*

$$(4.17) \qquad \mathrm{M}(\bar{R}_{\mathbf{l},\mathbf{q}}(q_n - l_m - n + m)) = \frac{1}{2}\,\mathrm{M}(\bar{R}_{\mathbf{l},\mathbf{q}^{(n)}}(q_n - l_m - n + m)).$$

PROOF. For $\mathbf{l} \neq \emptyset$, the lengths of the base and top side of $\bar{R}_{\mathbf{l},\mathbf{q}}(x)$ are x and $x + (l_m - m) - (q_n - n)$, respectively. Therefore, the assumption in (a) implies that the base is at most as long as the top side. Thus, by setting $x = 0$ we obtain a valid region, which has all the lozenges along its southwestern side forced (see Figure 4.8(a)). It is not hard to see that the leftover region is precisely $\bar{R}_{\mathbf{l}^{(m)},\mathbf{q}}(l_m - l_{m-1} - 1)$.

For part (b), note that our assumption implies that the top side is at most as long as the base. Indeed, we have already seen this for $\mathbf{l} \neq \emptyset$. For $\mathbf{l} = \emptyset$, the top side and the base have lengths $x - (q_n - n)$ and $x - q_1 + 1$, respectively. Therefore, the top side is as least as long as the base in this case as well.

By setting $x = q_n - l_m - n + m$, the top side shrinks to a point and we obtain a valid region whose lozenges along the northwestern side are forced (see Figure 4.8(b)). The leftover region is readily seen to be $\bar{R}_{\mathbf{l},\mathbf{q}^{(n)}}(q_n - l_m - n + m)$. Since the top forced tile has weight $1/2$ and all others have weight 1, this proves (4.17). $\quad\square$

5

Proof of Proposition 2.1

In this section we show that the polynomials $P_{\mathbf{l},\mathbf{q}}(x)$ and $\bar{P}_{\mathbf{l},\mathbf{q}}(x)$ satisfy the same recurrences as $\mathrm{M}(R_{\mathbf{l},\mathbf{q}}(x))$ and $\mathrm{M}(\bar{R}_{\mathbf{l},\mathbf{q}}(x))$. Then we deduce Proposition 2.1 by induction on $l_m + n + x$.

Rewrite formulas (1.5) and (1.6) as follows:

$$P_{\mathbf{l},\mathbf{q}}(x) = c_{\mathbf{l},\mathbf{q}} B_{m,n}(x + l_m - m) \prod_{i=1}^{m} \prod_{j=i}^{l_i-1} (x + l_m - j)(x + l_m - m + n + j + 2)$$

$$(5.1) \qquad \times \prod_{i=1}^{n} \prod_{j=i}^{q_i-1} (x + l_m - m + n - j + 1)(x + l_m + j + 1)$$

$$\bar{P}_{\mathbf{l},\mathbf{q}}(x) = \bar{c}_{\mathbf{l},\mathbf{q}} \bar{B}_{m,n}(x + l_m - m) \prod_{i=1}^{m} \prod_{j=i}^{l_i-1} (x + l_m - j)(x + l_m - m + n + j + 1)$$

$$(5.2) \qquad \times \prod_{i=1}^{n} \prod_{j=i}^{q_i-1} (x + l_m - m + n - j)(x + l_m + j + 1).$$

LEMMA 5.1. *For $m \le n$, we have*

$$(5.3) \qquad P_{\mathbf{l},\mathbf{q}}(x) = \sum_{k=1}^{n} (-1)^{n-k} C_k P_{\mathbf{l},\mathbf{q}^{(k)}}(x) + (-1)^n \delta_{mn} \bar{P}_{\mathbf{l},\mathbf{q}}(x),$$

where δ_{mn} is the Kronecker symbol and C_k is given by (4.3).

PROOF. The outline of our proof is the following. Dividing both sides of (5.3) by $C_n P_{\mathbf{l},\mathbf{q}}(x)$, the terms on the right hand side turn out to become ratios of linear polynomials in x, all having the same numerator. Dividing through further by this common numerator, the identity to be checked becomes the partial fraction decomposition of the rational function on the left hand side. Since the denominator of the latter is a product of distinct linear polynomials in x, this identity can be readily verified.

The expression on the right hand side of (4.3) giving C_k can be rewritten as

$$(5.4) \qquad C_k = \frac{(2x + 2l_m - m + n + 2)(x + l_m - q_k - m + n + 2)_{2q_k+m-n-1}}{2\,(2q_k + m - n)!}.$$

The left hand side. We have

135

$$\frac{P_{1,q}}{C_n P_{1,q^{(n)}}} = \frac{c_{1,q}}{c_{1,q^{(n)}}} \frac{2\,(2q_n + m - n)\,!}{(2x + 2l_m - m + n + 2)(x + l_m - q_n - m + n + 2)_{2q_n + m - n - 1}}$$

$$\times \frac{\prod_{i=1}^{m} \prod_{j=i}^{l_i - 1} (x + l_m - j)(x + l_m - m + n + j + 2)}{\prod_{i=1}^{m} \prod_{j=i}^{l_i - 1} (x + l_m - j)(x + l_m - m + n + j + 1)}$$

$$\times \frac{\prod_{i=1}^{n} \prod_{j=i}^{q_i - 1} (x + l_m + j + 1)(x + l_m - m + n - j + 1)}{\prod_{i=1}^{n-1} \prod_{j=i}^{q_i - 1} (x + l_m + j + 1)(x + l_m - m + n - j)}$$

$$\times \frac{B_{m,n}(x + l_m - m)}{B_{m,n-1}(x + l_m - m)}$$

$$= \frac{2^{\binom{n-m}{2} - m + 1}}{2^{\binom{n-m-1}{2} - m}} \, (2q_n + m - n)\,!$$

$$\times \frac{\prod_{i=1}^{m} \dfrac{1}{(2l_i)\,!} \prod_{i=1}^{n} \dfrac{1}{(2q_i - 1)\,!} \dfrac{\prod_{1 \le i < j \le m}(l_j - l_i) \prod_{1 \le i < j \le n}(q_j - q_i)}{\prod_{i=1}^{m} \prod_{j=1}^{n}(l_i + q_j)}}{\prod_{i=1}^{m} \dfrac{1}{(2l_i)\,!} \prod_{i=1}^{n-1} \dfrac{1}{(2q_i - 1)\,!} \dfrac{\prod_{1 \le i < j \le m}(l_j - l_i) \prod_{1 \le i < j \le n-1}(q_j - q_i)}{\prod_{i=1}^{m} \prod_{j=1}^{n-1}(l_i + q_j)}}$$

$$\times \prod_{i=1}^{m} \frac{x + l_m - m + n + l_i + 1}{x + l_m - m + n + i + 1} \prod_{i=1}^{n-1} \frac{x + l_m - m + n - i + 1}{x + l_m - m + n - q_i + 1}$$

$$\times \prod_{j=n}^{q_n - 1} (x + l_m + j + 1)(x + l_m - m + n - j + 1)$$

$$(5.5) \qquad \times \frac{1}{2^{m+n}}(2x + 2l_m - m + n + 2)_{m+n} \prod_{i=0}^{m-1} \frac{x + l_m + n - i + 1}{x + l_m + n - i + 1/2},$$

where we used the (easily checked) fact that

$$(5.6) \qquad \frac{B_{m,n}(x)}{B_{m,n-1}(x)} = \frac{1}{2^{m+n}}(2x + m + n + 2)_{m+n} \prod_{i=0}^{m-1} \frac{x + m + n - i + 1}{x + m + n - i + 1/2}.$$

After cancelling out common factors in the numerator and denominator of (5.5), and then rearranging the remaining factors, we obtain

$$\frac{P_{\mathbf{l},\mathbf{q}}}{C_n P_{\mathbf{l},\mathbf{q}^{(n)}}} = \frac{(2q_n + m - n)!}{(2q_n - 1)!} \frac{\prod_{i=1}^{n-1}(q_n - q_i)}{\prod_{i=1}^{m}(l_i + q_n)}$$

(5.7)
$$\times \frac{(2x + 2l_m - m + n + 2)_{n-m}}{(2x + 2l_m - m + n + 2)} \frac{\prod_{i=1}^{m}(x + l_m + l_i - m + n + 1)}{\prod_{i=1}^{n-1}(x + l_m - q_i - m + n + 1)}.$$

The terms in the sum on the right hand side. Using (5.1) and (5.2) we obtain

$$\frac{C_k P_{\mathbf{l},\mathbf{q}^{(k)}}}{C_n P_{\mathbf{l},\mathbf{q}^{(n)}}} = \frac{c_{\mathbf{l},\mathbf{q}^{(k)}}}{c_{\mathbf{l},\mathbf{q}^{(n)}}} \frac{(2q_n + m - n)!}{(2q_k + m - n)!} \frac{(x + l_m - q_k - m + n + 2)_{2q_k + m - n - 1}}{(x + l_m - q_n - m + n + 2)_{2q_n + m - n - 1}}$$

$$\times \frac{\prod_{i=k}^{n-1} \prod_{j=i}^{q_{i+1}-1}(x + l_m + j + 1)(x + l_m - m + n - j)}{\prod_{i=k}^{n-1} \prod_{j=i}^{q_i - 1}(x + l_m + j + 1)(x + l_m - m + n - j)}$$

$$= \frac{(2q_n + m - n)!}{(2q_k + m - n)!}$$

$$\times \frac{\prod_{i=1}^{m} \frac{1}{(2l_i)!} \prod_{i=1, i\neq k}^{n} \frac{1}{(2q_i - 1)!} \frac{\prod_{1\le i<j\le m}(l_j - l_i) \prod_{1\le i<j\le n, i,j\neq k}(q_j - q_i)}{\prod_{i=1}^{m} \prod_{j=1, j\neq k}^{n}(l_i + q_j)}}{\prod_{i=1}^{m} \frac{1}{(2l_i)!} \prod_{i=1}^{n-1} \frac{1}{(2q_i - 1)!} \frac{\prod_{1\le i<j\le m}(l_j - l_i) \prod_{1\le i<j\le n-1}(q_j - q_i)}{\prod_{i=1}^{m} \prod_{j=1}^{n-1}(l_i + q_j)}}$$

$$\times \frac{(x + l_m - q_k - m + n + 2)_{2q_k + m - n - 1}}{(x + l_m - q_n - m + n + 2)_{2q_n + m - n - 1}}$$

(5.8)
$$\times \prod_{i=k}^{n-1} \prod_{j=q_i}^{q_{i+1}-1}(x + l_m + j + 1)(x + l_m - m + n - j).$$

By cancelling out common factors and regrouping one obtains

$$\frac{C_k P_{\mathbf{l},\mathbf{q}^{(k)}}}{C_n P_{\mathbf{l},\mathbf{q}^{(n)}}} = \frac{(2q_k - 1)!}{(2q_n - 1)!} \frac{(2q_n + m - n)!}{(2q_k + m - n)!} \prod_{i=1}^{m} \frac{l_i + q_k}{l_i + q_n}$$

(5.9)
$$\times \frac{\prod_{i=1, i\neq k}^{n-1}(q_n - q_i)}{\prod_{i=1}^{k-1}(q_k - q_i) \prod_{i=k+1}^{n-1}(q_i - q_k)} \frac{x + l_m - q_n - m + n + 1}{x + l_m - q_k - m + n + 1}.$$

The last term on the right hand side. Since for $m \neq n$ this term is 0, assume $m = n$. We have

$$\frac{\bar{P}_{1,\mathbf{q}}}{C_n P_{1,\mathbf{q}^{(n)}}} = \frac{\bar{c}_{1,\mathbf{q}}}{c_{1,\mathbf{q}^{(n)}}} \frac{2\,(2q_n)\,!}{(2x+2l_n+2)(x+l_n-q_n+2)_{2q_n-1}}$$

$$\times \frac{\prod_{i=1}^{n}\prod_{j=i}^{l_i-1}(x+l_n-j)(x+l_n+j+1)}{\prod_{i=1}^{n}\prod_{j=i}^{l_i-1}(x+l_n-j)(x+l_n+j+1)}$$

$$\times \frac{\prod_{i=1}^{n}\prod_{j=i}^{q_i-1}(x+l_n-j)(x+l_n+j+1)}{\prod_{i=1}^{n-1}\prod_{j=i}^{q_i-1}(x+l_n-j)(x+l_n+j+1)}$$

$$\times \frac{\bar{B}_{n,n}(x+l_n-n)}{B_{n,n-1}(x+l_n-n)}$$

$$= \frac{2^{\binom{0}{2}-n+1}}{2^{\binom{-1}{2}-n}}\,(2q_n)\,!$$

$$\times \frac{\prod_{i=1}^{n}\dfrac{1}{(2l_i-1)\,!}\prod_{i=1}^{n}\dfrac{1}{(2q_i)\,!}\dfrac{\prod_{1\le i<j\le n}(l_j-l_i)\prod_{1\le i<j\le n}(q_j-q_i)}{\prod_{i=1}^{n}\prod_{j=1}^{n}(l_i+q_j)}}{\prod_{i=1}^{n}\dfrac{1}{(2l_i)\,!}\prod_{i=1}^{n-1}\dfrac{1}{(2q_i-1)\,!}\dfrac{\prod_{1\le i<j\le n}(l_j-l_i)\prod_{1\le i<j\le n-1}(q_j-q_i)}{\prod_{i=1}^{n}\prod_{j=1}^{n-1}(l_i+q_j)}}$$

$$\times \frac{\prod_{j=n}^{q_n-1}(x+l_n-j)(x+l_n+j+1)}{(2x+2l_n+2)(x+l_n-q_n+2)_{2q_n-1}}$$

(5.10)
$$\times (x+l_n-n+1)_{2n},$$

where we have used the equality

(5.11)
$$\frac{\bar{B}_{n,n}(x)}{B_{n,n-1}(x)} = (x+1)_{2n},$$

which can be readily checked. Simplifying and rearranging in (5.10) leads to

(5.12)
$$\frac{\bar{P}_{1,\mathbf{q}}}{C_n P_{1,\mathbf{q}^{(n)}}} = \frac{l_1\cdots l_n}{q_1\cdots q_{n-1}}\frac{\prod_{i=1}^{n-1}(q_n-q_i)}{\prod_{i=1}^{n}(l_i+q_n)}\frac{x+l_n-q_n+1}{x+l_n+1}.$$

By (5.7), (5.9) and (5.12), it follows that the statement of the Lemma is equivalent to the following rational function identity:

$$\frac{(2q_n + m - n)!}{(2q_n - 1)!} \frac{\prod_{i=1}^{n-1}(q_n - q_i)}{\prod_{i=1}^{m}(l_i + q_n)}$$

$$\times \frac{(2x + 2l_m - m + n + 2)_{n-m}}{(2x + 2l_m - m + n + 2)} \frac{\prod_{i=1}^{m}(x + l_m + l_i - m + n + 1)}{\prod_{i=1}^{n}(x + l_m - q_i - m + n + 1)}$$

(5.13)
$$= \sum_{k=1}^{n}(-1)^{n-k}\frac{\alpha_k}{x + l_m - q_k - m + n + 1} + \delta_{mn}\frac{\beta}{x + l_n - q_n + 1},$$

where α_k and β are given by

$$\alpha_k = \frac{(2q_k - 1)!\,(2q_n + m - n)!}{(2q_n - 1)!\,(2q_k + m - n)!}\frac{\prod_{i=1}^{m}(l_i + q_k)}{\prod_{i=1}^{m}(l_i + q_n)}\frac{\prod_{i=1,\,i\neq k}^{n-1}(q_n - q_i)}{\prod_{i=1}^{k-1}(q_k - q_i)\prod_{i=k+1}^{n-1}(q_i - q_k)}$$

$$\beta = \frac{l_1 \cdots l_n}{q_1 \cdots q_{n-1}}\frac{\prod_{i=1}^{n-1}(q_n - q_i)}{\prod_{i=1}^{n}(l_i + q_n)}.$$

However, (5.13) can be regarded as the partial fraction decomposition of the rational function on its left hand side. Since the denominator is factored in distinct linear factors, the values of the constants α_k and β that make (5.13) true can be immediately obtained by suitable specializations of x. It is easy to see that these values are precisely the ones in the above two equalities.   □

LEMMA 5.2. *Let D_k, $\bar{C}_k$ and $\bar{D}_k$ be given by* (4.7), (4.11) *and* (4.13).
(a) *For $m > n$, we have*

(5.14) $$P_{\mathbf{l},\mathbf{q}}(x) = \sum_{k=1}^{m-1}(-1)^{m-k}D_k P_{\mathbf{l}^{(k)},\mathbf{q}}(x - 1) + D_m P_{\mathbf{l}^{(m)},\mathbf{q}}(x + l_m - l_{m-1} - 1).$$

(b) *For $m < n$, we have*

(5.15) $$\bar{P}_{\mathbf{l},\mathbf{q}}(x) = \sum_{k=1}^{n}(-1)^{n-k}\bar{C}_k \bar{P}_{\mathbf{l},\mathbf{q}^{(k)}}(x).$$

(c) *For $m \geq n$, we have*

$$\bar{P}_{\mathbf{l},\mathbf{q}}(x) = \sum_{k=1}^{m-1}(-1)^{m-k}\bar{D}_k \bar{P}_{\mathbf{l}^{(k)},\mathbf{q}}(x - 1) + \bar{D}_m \bar{P}_{\mathbf{l}^{(m)},\mathbf{q}}(x + l_m - l_{m-1} - 1)$$

(5.16) $$+ (-1)^m \delta_{mn} P_{\mathbf{l},\mathbf{q}}(x - 1).$$

PROOF. The identities (5.14)–(5.16) can be proved using the same approach as in the proof of Lemma 5.1. For the first two, we divide through by the last term on the right hand side, and for the third by the second to last. In each of the three cases, the terms on the right hand sides become ratios of linear polynomials, all with the same numerator. Dividing through by this common numerator, the identity to be proved becomes the partial fraction decomposition of the rational function on the left hand side. In all three cases, the denominator of this rational function is a product of distinct linear factors, and thus, as in the proof of Lemma 5.1, the identity is easily checked by suitable specializations of x.

This can be carried out in a straightforward way. The shifts in the arguments in (5.14)–(5.16) are so that the division mentioned above results in great simplification of the involved expressions. The only difference that needs to be pointed out is that we need the following relations, analogous to (5.6) and (5.11), involving the monic polynomials $B_{m,n}$ and $\bar{B}_{m,n}$:

$$\frac{B_{m,n}(x)}{B_{m-1,n}(x)} = \frac{1}{2^{m+n-1}}(x+m+n)(x+m+n+1)(2x+m+n+2)_{m+n-1}$$

$$\times \prod_{i=0}^{n-1} \frac{x+m+i}{x+m+i+1/2}$$

$$\frac{\bar{B}_{m,n}(x)}{\bar{B}_{m,n-1}(x)} = \frac{1}{2^{m+n}}(x+m+n)(2x+m+n+1)_{m+n}$$

$$\times \prod_{i=0}^{m-1} \frac{x+m+n-i-1}{x+m+n-i-1/2}$$

$$\frac{\bar{B}_{m,n}(x)}{\bar{B}_{m-1,n}(x)} = \frac{1}{2^{m+n}}(2x+m+n+1)_{m+n} \prod_{i=0}^{n-1} \frac{x+m+i+1}{x+m+i+1/2}$$

$$\frac{B_{n,n}(x-1)}{\bar{B}_{n-1,n}(x)} = \frac{(x)_{2n+1}}{x+n}.$$

These are all easily verified directly using the defining formulas for $B_{m,n}$ and $\bar{B}_{m,n}$.   □

LEMMA 5.3. *We have*

$$P_{\mathbf{l},\mathbf{q}}(q_n - l_m - n + m - 1) = \frac{1}{2}P_{\mathbf{l},\mathbf{q}^{(n)}}(q_n - l_m - n + m - 1)$$

$$\bar{P}_{\mathbf{l},\mathbf{q}}(q_n - l_m - n + m) = \frac{1}{2}\bar{P}_{\mathbf{l},\mathbf{q}^{(n)}}(q_n - l_m - n + m),$$

and for $\mathbf{l} \neq \emptyset$ *we have*

$$P_{\mathbf{l},\mathbf{q}}(0) = P_{\mathbf{l}^{(m)},\mathbf{q}}(l_m - l_{m-1} - 1)$$

$$\bar{P}_{\mathbf{l},\mathbf{q}}(0) = \bar{P}_{\mathbf{l}^{(m)},\mathbf{q}}(l_m - l_{m-1} - 1).$$

PROOF. For $m \leq n$, from (5.7) and (5.4) one obtains a simple expression for the ratio $P_{\mathbf{l},\mathbf{q}}(x)/P_{\mathbf{l},\mathbf{q}^{(n)}}(x)$. It is readily seen that for $x = q_n - l_m - n + m - 1$ this expression becomes $1/2$, thus proving the first equality in the statement of the Lemma in this case. For $m > n$, one obtains from (5.5) an expression for $P_{\mathbf{l},\mathbf{q}}(x)/(C_n P_{\mathbf{l},\mathbf{q}^{(n)}}(x))$ similar to (5.7). Again, it follows easily that the ratio $P_{\mathbf{l},\mathbf{q}}(x)/P_{\mathbf{l},\mathbf{q}^{(n)}}(x)$ specializes to $1/2$ for $x = q_n - l_m - n + m - 1$.

The remaining three equalities are proved similarly, just as the proof of Lemma 5.2 is similar to that of Lemma 5.1.   □

PROOF OF PROPOSITION 2.1. We proceed by induction on $l_m + n + x$. For $R_{\mathbf{l},\mathbf{q}}(x)$, the minimum possible value of $l_m + n + x$ is -1. In this case $\mathbf{l}$ and $\mathbf{q}$ are necessarily empty, and (2.1) is true by our definitions. For $\bar{R}_{\mathbf{l},\mathbf{q}}(x)$, the minimum value of $l_m + n + x$ is 0. Again, this implies $\mathbf{l} = \mathbf{q} = \emptyset$, and (2.2) follows from our definitions.

Let $N > 0$ and suppose that (2.1) and (2.2) are true for all $\mathbf{l}$, $\mathbf{q}$ and x with $l_m + n + x < N$. Consider $\mathbf{l}$, $\mathbf{q}$ and x so that $l_m + n + x = N$. We claim that (2.2) holds also for these choices of $\mathbf{l}$, $\mathbf{q}$ and x.

Indeed, assume first that x is not equal to its minimum possible value. Suppose $m \geq n$. By (4.12), $M(\bar{R}_{\mathbf{l},\mathbf{q}}(x))$ can be expressed in terms of tiling generating functions of regions from the $\bar{R}$- and (in case $m = n$) R-family. By the induction hypothesis, (2.1) and (2.2) hold for all the regions on the right hand side of (4.12). Then Lemma 5.2(c) implies that (2.2) holds in this case. For $m < n$, (2.2) follows similarly, from (4.10), the induction hypothesis and Lemma 5.2(b).

Assume now that x is equal to its minimum possible value. Then x equals the value of the argument on the left hand side of either (4.16) or (4.17), and we can express $M(\bar{R}_{\mathbf{l},\mathbf{q}})(x)$ in terms of the tiling generating function of a region from the $\bar{R}$-family for which the induction hypothesis applies. Therefore, Lemma 5.3 implies that (2.2) holds for $\bar{R}_{\mathbf{l},\mathbf{q}}(x)$ as well.

We now show that (2.1) holds for $l_m + n + x = N$. Assume that x is not equal to its minimum possible value. Suppose $m \leq n$, and apply Lemma 4.2. By the induction hypothesis, (2.1) holds for the regions appearing in the terms of the sum on the right hand side of (4.2). Furthermore, we have seen above that (2.2) holds for $\bar{R}_{\mathbf{l},\mathbf{q}}(x)$. Therefore, by Lemma 5.1 we obtain that (2.1) is true for $R_{\mathbf{l},\mathbf{q}}(x)$, if $m \leq n$. The case $m > n$ follows analogously, using Lemma 4.3, the induction hypothesis and Lemma 5.2(a).

Finally, if x takes on its minimum possible value, (2.1) follows by (4.14) and (4.15), the induction hypothesis and Lemma 5.3. This completes the proof of Proposition 2.1 by induction. $\qquad \square$

The guessing of $\mathrm{M}(R_{\mathbf{l},\mathbf{q}}(x))$ and $\mathrm{M}(\bar{R}_{\mathbf{l},\mathbf{q}}(x))$

In this section we describe how we arrived at conjecturing that the tiling generating functions for the regions $R_{\mathbf{l},\mathbf{q}}(x)$ and $\bar{R}_{\mathbf{l},\mathbf{q}}(x))$ are given by the polynomials $P_{\mathbf{l},\mathbf{q}}(x)$ and $\bar{P}_{\mathbf{l},\mathbf{q}}(x)$.

We discuss here the case of $\bar{R}_{\mathbf{l},\mathbf{q}}(x)$. The region $R_{\mathbf{l},\mathbf{q}}(x)$ can be treated similarly.

It is easy to write down explicitly the entries of the matrix in (4.1). This gives an easy way to compute the polynomials $\mathrm{M}(\bar{R}_{\mathbf{l},\mathbf{q}}(x))$ by computer, for specific choices of $\mathbf{l}$ and $\mathbf{q}$. At once, one is striked by the fact that all these polynomials seem to factor in the form

$$(6.1) \qquad \mathrm{M}(\bar{R}_{\mathbf{l},\mathbf{q}}(x)) = \bar{c}_{\mathbf{l},\mathbf{q}} F_{\mathbf{l},\mathbf{q}}(x),$$

where $\bar{c}_{\mathbf{l},\mathbf{q}}$ is a constant and $F_{\mathbf{l},\mathbf{q}}(x)$ is the product of linear factors of the form $(x + t)$, where t is either an integer or a half-integer.

Moreover, it is easy to conjecture the behaviour of $F_{\mathbf{l},\mathbf{q}}(x)$ under incrementing a single element of $\mathbf{l}$ or $\mathbf{q}$. Specifically, after some experimenting one soon comes up with the guess that

$$(6.2)$$
$$\frac{F_{\mathbf{l}^{|k\rangle},\mathbf{q}}(x)}{F_{\mathbf{l},\mathbf{q}}(x)} = (x - l_k + l_m)(x + l_k + l_m - m + n + 1), \qquad \text{for } 1 \le k < m$$

$$(6.3)$$
$$\frac{F_{\mathbf{l}^{|m\rangle},\mathbf{q}}(x - 1)}{F_{\mathbf{l},\mathbf{q}}(x)} = x(x + 2l_m - m + n + 1)$$

$$(6.4)$$
$$\frac{F_{\mathbf{l},\mathbf{q}^{|k\rangle}}(x)}{F_{\mathbf{l},\mathbf{q}}(x)} = (x + q_k + l_m + 1)(x - q_k + l_m - m + n), \qquad \text{for } 1 \le k \le n,$$

where $\mathbf{l}^{|k\rangle}$ is the list obtained from $\mathbf{l}$ by increasing its k-th element by 1 (so for $k < m$, $\mathbf{l}^{|k\rangle}$ is defined only if $l_{k+1} - l_k \ge 2$). These formulas provide us with an explicit guess for $F_{\mathbf{l},\mathbf{q}}$, in terms of $F_{[m],[n]}(x)$, where $[m]$ is the list consisting of $1, \ldots, m$. Furthermore, some experimentation easily leads one to conjecture that $F_{[m],[n]}(x) = \bar{B}_{m,n}(x)$ (where $\bar{B}_{m,n}(x)$ is given by (1.2)).

It is considerably more difficult to guess a formula for the constants $\bar{c}_{\mathbf{l},\mathbf{q}}$ just based on working out examples (this is not surprising, as factorization of integers gives significantly less information than factorization of polynomials). However, it turns out that we do not need to guess the value of these constants: substituting the conjectured expression for $F_{\mathbf{l},\mathbf{q}}$ in (6.1) and using then Lemma 4.7, we obtain (conjectured) recurrences satisfied by the $\bar{c}_{\mathbf{l},\mathbf{q}}$'s.

More precisely, suppose $l_m - m \geq q_n - n$. We may then use (4.16) to relate $\bar{c}_{l,q}$ to $\bar{c}_{l^{(m)},q}$. Carrying this out one arrives at

$$(6.5) \qquad \frac{\bar{c}_{l,q}}{\bar{c}_{l^{(m)},q}} = 2^{m-n-1} \frac{1}{(2l_m-1)!} \frac{\prod_{i=1}^{m-1}(l_m - l_i)}{\prod_{i=1}^{n}(l_m + q_i)}.$$

On the other hand, if $l_m - m \leq q_n - n$, we deduce similarly from (4.17) that

$$(6.6) \qquad \frac{\bar{c}_{l,q}}{\bar{c}_{l,q^{(n)}}} = 2^{n-m-1} \frac{1}{(2q_n)!} \frac{\prod_{i=1}^{n-1}(q_n - q_i)}{\prod_{i=1}^{m}(q_n + l_i)}.$$

We can then repeat this procedure for l and q replaced by $l^{(m)}$ and q or l and $q^{(n)}$, depending on which of (4.16) and (4.17) applies. At each iteration we have to look at the initial segments of the lists l and q that are left over, and compare the difference between their largest entry and their number of entries, before deciding which of (6.5) or (6.6) to apply. However, it is not hard to see that no matter in which order we apply these recurrences, they lead to an expression of the form

$$\bar{c}_{l,q} = 2^{e(l,q)} \prod_{i=1}^{m} \frac{1}{(2l_i-1)!} \prod_{i=1}^{n} \frac{1}{(2q_i)!} \frac{\prod_{1 \leq i < j \leq m}(l_j - l_i) \prod_{1 \leq i < j \leq n}(q_j - q_i)}{\prod_{i=1}^{m}\prod_{j=1}^{n}(l_i + q_j)},$$

where $e(l,q)$ is some integer depending only on l and q. Then after some experimenting one is readily lead to guess that $e(l,q) = \binom{n-m}{2} - m$, and we obtain (1.4). It is remarkable how the "weak" conjectures (6.1) and (6.2)–(6.4) lead us, by the "forcing" argument in Lemma 4.7, to the precise (conjectured) formula for $\bar{c}_{l,q}$.

A similar analysis can be done for $M(R_{l,q}(x))$, and one arrives at formulas (1.5), (1.1) and (1.3).

Acknowledgments. David Wilson's program Vaxmacs for counting perfect matchings was very useful in finding the families of regions $R_{l,q}(x)$ and $\bar{R}_{l,q}(x)$.

Bibliography

[1] M. Ciucu, *Enumeration of perfect matchings in graphs with reflective symmetry*, J. Comb. Theory Ser. A **77** (1997), 67–97.

[2] M. Ciucu, *Enumeration of lozenge tilings of punctured hexagons*, J. Combin. Theory Ser. A **83** (1998), 268–272.

[3] G. David and C. Tomei, *The problem of the calissons*, Amer. Math. Monthly **96** (1989), 429–431.

[4] S. Elnitsky, *Rhombic tilings of polygons and classes of reduced words in Coxeter groups*, J. Combin. Theory Ser. A **77** (1997), 193–221.

[5] I. M. Gessel and X. Viennot, *Binomial determinants, paths, and hook length formulae*, Adv. in Math. **58** (1985), 300–321.

[6] C. Krattenthaler, *Generating functions for plane partitions of a given shape*, Manuscripta Math. **69** (1990), 173–201.

[7] G. Kuperberg, *Symmetries of plane partitions and the permanent-determinant method*, J. Combin. Theory Ser. A **68** (1994), 115–151.

[8] P. A. MacMahon, *Memoir on the theory of the partition of numbers—Part V. Partitions in two-dimensional space*, Phil. Trans. R. S. (1911), A.

[9] P. A. MacMahon, *Combinatory Analysis, vols. 1–2*, Chelsea, New York, 1960.

[10] J. Propp, *Enumeration of matchings: problems and progress, in "New Perspectives in Geometric Combinatorics"*, MSRI Publications **38** (1999), 255–291.

[11] R. P. Stanley, *Ordered structures and partitions*, Memoirs of the Amer. Math. Soc. **119** (1972).

[12] J. R. Stembridge, *Nonintersecting paths, Pfaffians and plane partitions*, Adv. in Math. **83** (1990), 96–131.

Editorial Information

To be published in the *Memoirs*, a paper must be correct, new, nontrivial, and significant. Further, it must be well written and of interest to a substantial number of mathematicians. Piecemeal results, such as an inconclusive step toward an unproved major theorem or a minor variation on a known result, are in general not acceptable for publication. Papers appearing in *Memoirs* are generally at least 80 and not more than 200 published pages in length. Papers less than 80 or more than 200 published pages require the approval of the Managing Editor of the Transactions/Memoirs Editorial Board.

As of July 31, 2005, the backlog for this journal was approximately 14 volumes. This estimate is the result of dividing the number of manuscripts for this journal in the Providence office that have not yet gone to the printer on the above date by the average number of monographs per volume over the previous twelve months, reduced by the number of volumes published in four months (the time necessary for preparing a volume for the printer). (There are 6 volumes per year, each containing at least 4 numbers.)

A Consent to Publish and Copyright Agreement is required before a paper will be published in the *Memoirs*. After a paper is accepted for publication, the Providence office will send a Consent to Publish and Copyright Agreement to all authors of the paper. By submitting a paper to the *Memoirs*, authors certify that the results have not been submitted to nor are they under consideration for publication by another journal, conference proceedings, or similar publication.

Information for Authors

Memoirs are printed from camera copy fully prepared by the author. This means that the finished book will look exactly like the copy submitted.

The paper must contain a *descriptive title* and an *abstract* that summarizes the article in language suitable for workers in the general field (algebra, analysis, etc.). The *descriptive title* should be short, but informative; useless or vague phrases such as "some remarks about" or "concerning" should be avoided. The *abstract* should be at least one complete sentence, and at most 300 words. Included with the footnotes to the paper should be the 2000 *Mathematics Subject Classification* representing the primary and secondary subjects of the article. The classifications are accessible from www.ams.org/msc/. The list of classifications is also available in print starting with the 1999 annual index of *Mathematical Reviews*. The Mathematics Subject Classification footnote may be followed by a list of *key words and phrases* describing the subject matter of the article and taken from it. Journal abbreviations used in bibliographies are listed in the latest *Mathematical Reviews* annual index. The series abbreviations are also accessible from www.ams.org/publications/. To help in preparing and verifying references, the AMS offers MR Lookup, a Reference Tool for Linking, at www.ams.org/mrlookup/. When the manuscript is submitted, authors should supply the editor with electronic addresses if available. These will be printed after the postal address at the end of the article.

Electronically prepared manuscripts. The AMS encourages electronically prepared manuscripts, with a strong preference for $\mathcal{AMS}$-LaTeX. To this end, the Society has prepared $\mathcal{AMS}$-LaTeX author packages for each AMS publication. Author packages include instructions for preparing electronic manuscripts, the *AMS Author Handbook*, samples, and a style file that generates the particular design specifications of that publication series. Though $\mathcal{AMS}$-LaTeX is the highly preferred format of TeX, author packages are also available in $\mathcal{AMS}$-TeX.

Authors may retrieve an author package from e-MATH starting from www.ams.org/tex/ or via FTP to ftp.ams.org (login as anonymous, enter username as password, and type cd pub/author-info). The *AMS Author Handbook* and the *Instruction Manual* are available in PDF format following the author packages link from www.ams.org/tex/. The author package can be obtained free of charge by sending email to pub@ams.org (Internet) or from the Publication Division, American Mathematical Society, 201 Charles St., Providence, RI 02904, USA. When requesting an author package, please specify $\mathcal{AMS}$-LATEX or $\mathcal{AMS}$-TEX, Macintosh or IBM (3.5) format, and the publication in which your paper will appear. Please be sure to include your complete mailing address.

Sending electronic files. After acceptance, the source file(s) should be sent to the Providence office (this includes any TEX source file, any graphics files, and the DVI or PostScript file).

Before sending the source file, be sure you have proofread your paper carefully. The files you send must be the EXACT files used to generate the proof copy that was accepted for publication. For all publications, authors are required to send a printed copy of their paper, which exactly matches the copy approved for publication, along with any graphics that will appear in the paper.

TEX files may be submitted by email, FTP, or on diskette. The DVI file(s) and PostScript files should be submitted only by FTP or on diskette unless they are encoded properly to submit through email. (DVI files are binary and PostScript files tend to be very large.)

Electronically prepared manuscripts can be sent via email to pub-submit@ams.org (Internet). The subject line of the message should include the publication code to identify it as a Memoir. TEX source files, DVI files, and PostScript files can be transferred over the Internet by FTP to the Internet node e-math.ams.org (130.44.1.100).

Electronic graphics. Comprehensive instructions on preparing graphics are available at www.ams.org/jourhtml/graphics.html. A few of the major requirements are given here.

Submit files for graphics as EPS (Encapsulated PostScript) files. This includes graphics originated via a graphics application as well as scanned photographs or other computer-generated images. If this is not possible, TIFF files are acceptable as long as they can be opened in Adobe Photoshop or Illustrator. No matter what method was used to produce the graphic, it is necessary to provide a paper copy to the AMS.

Authors using graphics packages for the creation of electronic art should also avoid the use of any lines thinner than 0.5 points in width. Many graphics packages allow the user to specify a "hairline" for a very thin line. Hairlines often look acceptable when proofed on a typical laser printer. However, when produced on a high-resolution laser imagesetter, hairlines become nearly invisible and will be lost entirely in the final printing process.

Screens should be set to values between 15% and 85%. Screens which fall outside of this range are too light or too dark to print correctly. Variations of screens within a graphic should be no less than 10%.

Inquiries. Any inquiries concerning a paper that has been accepted for publication should be sent directly to the Electronic Prepress Department, American Mathematical Society, 201 Charles St., Providence, RI 02904, USA.

Titles in This Series

TITLES IN THIS SERIES

For a complete list of titles in this series, visit the
AMS Bookstore at **www.ams.org/bookstore/**.